And secondly...

Welcome, dear reader, or possibly 'welcome back', as I'm sort of assuming you will already have read *Scenes from a Smallholding*, to which *More Scenes* is a sequel. Actually, I did think of calling this volume *Son of Scenes from a Smallholding*, but the editor thought that would be silly, so we didn't.

That's about it, really.

- If you enjoyed *Scenes*, I dare to hope that you will enjoy *More Scenes* as well.
- If you *didn't* enjoy *Scenes* – well, what on earth are you doing reading this?
- If you've never *heard* of *Scenes* – well you have now. But be assured that you don't need to have read it in order to enjoy *More Scenes*.

I wish you joy and laughter.

Have a wonderful day!

Chas Griffin

* * * * * * *

A thousand times a million men
have hoed a billion weeds
but I have yet to hear of one
who hasn't once felt wonder at

Forthcoming titles

Following the enormous success of
Seventy-two Ways of Wasting Money Quickly

the author is currently working on
Son of Seventy-two Ways of Wasting Money Quickly

and is also developing
Lose a Stone in Twenty Minutes!
Teach Yourself to Read

and a new edition of the evergreen
So You've Bought a Sinclair C5…

More Scenes from a Smallholding

CHAS GRIFFIN

EBURY PRESS

First published in Great Britain 2006

1 3 5 7 9 10 8 6 4 2

Ebury Press, an imprint of Ebury Publishing.
Random House, 20 Vauxhall Bridge Road, London SW1V 2SA

Random House Australia (Pty) Limited
20 Alfred Street, Milsons Point, Sydney, New South Wales 2061, Australia

Random House New Zealand Limited
18 Poland Road, Glenfield, Auckland 10, New Zealand

Random House (Pty) Limited
Isle of Houghton, Corner of Boundary Road & Carse O'Gowrie,
Houghton 2198, South Africa

Random House Publishers India Private Limited
301 World Trade Tower, Hotel Intercontinental Grand Complex,
Barakhamba Lane, New Delhi 110 001, India

The Random House Group Limited Reg. No. 954009

www.randomhouse.co.uk

A CIP catalogue record for this book is available from the British Library.

Cover design by Two Assocaites
Interior by seagulls.net

ISBN 9780091905880 (from Jan 2007)
ISBN 0091905885

Papers used by Eb recyclable products
made from nable forests.

Printed and bound in Great I Wyman Ltd, Reading, Berkshire

No goats were seri riting of this book.

Acknowledgements

Apart from the good people at Ebury … and Ken and Doreen next door … and Ken Guy … many thanks to Mark Stanton, my trusty agent; Scott Pack of Waterstone's for his faith in and support of *Scenes*; John Sheffield for his constant help and thoughtful crits; Nevil Hutchinson and Matthew Young, for their specialist input; and Jean Viggers and Alan Preston, for helping my memory along.

But mainly thanks to Anne, as ever. Thanks, love, for your endless kindness and patience.

And a very big thanks also to April for being the heart of the farm for two full decades. She never quite made it to her twenty-first, which was a pity because we had a big party planned, with lots of carrots and cabbage leaves and everything. Bye, sweetheart.

Dedication

To all my family,
plus dear old April (1985–2005)

Contents

Introduction and Caution

Well now, where was I? ... Ah yes, Christmas 1985. I was out for the count, in bed, smelling alarmingly of vinegar, with what had just been diagnosed as ME.* I was very slowly beginning to get the hang of the rules of snooker through watching endless tournaments on a tiny black and white telly, which sported more snow than a Swedish winter. That was about my intellectual limit.

I couldn't concentrate on anything as demanding as reading, let alone making a decision of any sort. My emotions were all over the place. I had no will. And, obviously, my body was as weak as several kittens.

This was not the ideal condition in which to greet our fourth season as in-at-the-deep-end organic smallholders.

Our previous three seasons had seen us advance from pig-ignorant townies to pig-wise and semi-competent growers of many types of veg. We also had a couple of polytunnels, a house cow, a bevy of amiable sheep, a pig or two, obviously, and the cutest little tractor you ever did see. We had a rotavator, a brand new orchard, and more poultry than you could shake a stick at, should you ever have the time, which, frankly, we did not. Busy, busy, busy ...

We were on the up, and pretty pleased with ourselves, and rightly so. We'd learned an amazing amount in three

---

* Or Post Viral Fatigue Syndrome; or Royal Free Disease; or Chronic Fatigue Syndrome; or, God help us... 'Yuppie Flu'.

years, not least that garlic isn't the very best cash crop to bank on in Wet West Wales. Nice try, though.*

Everything was looking rosy…

Then wallop: one evening, while I was sorting out a bit of Swiss Chard for Tim the wholesaler to take to Swansea the next morning, I suddenly faded. The flu I'd been nurturing for the last few days caught up with me like a runaway elephant hitting the buffers. Within three or four minutes I became incompetent and feeble. Within ten I was collapsed in my favourite prolapsed old armchair. And within the hour I was in bed, utterly exhausted and drifting in and out of a ponderous and unrefreshing sleep.

Who'd be a smallholder, eh? All that fresh air, healthy exercise and wonderful organic food – and where does it get you? Starfished under the duvet like a knife-thrower's target, wishing he'd get a move on.

The irony of it all *did* creep up on me actually, as I swilled in and out of what seemed to be passing for consciousness.

It was so baffling. Why had I 'caught' this totally debilitating condition? Surely I was in good shape, wasn't I? Better than I had been in suburbia? Fulfilled, creative, involved, happy, productive, healthy in body, mind and spirit? Yes, to all these. So…*why*?**

And that was Christmas '85.

---

* If you're a big fan of disasters you can find more details on the Great Garlic Fiasco, and our first three seasons in general, in *Scenes from a Smallholding*. Whenever you see this: (StaS) in the following text it means there's a bit more about the subject of the moment in *Scenes*.

** I'm pleased to recount that battered though I was, I didn't sink to 'Why *me*?', as it seemed that 'Why *not*?' was a perfectly adequate response.

So now, gentle reader, please read on, unless the cat needs letting out, in which case, better safe than sorry, I guess.

Chas Griffin, Newcastle Emlyn, West Wales
Spring 2006

Oh yes, I nearly forgot. If you've already read *Scenes from a Smallholding* then you won't be surprised by the rather eccentric layout of *More Scenes.* I've stuck more or less to the same pattern: an article on a topic, originally printed in the HDRA* magazine, *The Organic Way*, followed by some matters arising, and then a third section, offering some sort of loose story line.

Please don't get worried by the fact that you can't pin down a precise sequence of events, or that the storyline might even seem to dodge about from time to time. It really doesn't matter. What I'm trying to convey is a general *impression* of our life on the farm: triumphs, warts, hysteria, fun, disasters and all. And the articles were often written years after the events they describe. Some were written outside the timeframe of the book. Everything I write is true, however, except for the really silly bits. No, I did not 'hobble back to the cellar, wild-eyed and cackling', and nor did I then 'throw the mighty switch'. For a start we don't have a cellar, *or* a mighty switch. Nor am I given to hobbling and cackling. I'd have thought that much was obvious. Just use your own judgement.

And please don't worry that some measurements are given

* The Henry Doubleday Research Association, the largest organic gardening society in Europe, with 30,000+ members; HDRA has recently changed its name to *Garden Organic.*

in metric and others in imperial. We're in transition still, and some things are currently best dealt with using one system, and others using the other, it seems to me. (I once had a student hand in an essay in which he told me a table-tennis table measured 2743mm by 1524mm. No no no … it measures 9ft x 5ft.)

Just for the record, *Scenes* was concerned with why and how we left suburbia for the wilds, and followed our progress from our arrival to the day I got ill. As you will already have gathered, *More Scenes* takes up where *Scenes* left off, and stumbles on until ... well, I'll leave you to find out.

And yes, dour Nemesis may well make another cameo appearance, just in case you were wondering.

Plans

We'd already made a lot of tentative plans before we arrived in Wales. We knew that we wanted to set up and run a closed organic system: that is, one which would buy in as little as possible, and export as little as necessary. Our prime aim was to improve the land as much as we could, and to grow superb crops for ourselves, and perhaps for a few neighbours. We wanted to leave a little corner of the world happier for our having been there.

Does this sound pretentious to you? I hope not, because it seems to me that if everyone had similar plans to take care of the land, the world would be in a far happier condition than it is. We would not be losing our topsoil at the rate of millions of tons a month, as erosion swills and blows it away: the simple result of having ruined the soil by bad farming practice, driven by short-term profit rather than the absolute need for long-term sustainability. Cause: effect. Wind: whirlwind. Very, very simple, wouldn't you say? But either too difficult for our super-intelligent leaders to comprehend, or maybe they just don't care. Maybe if your snout's in the trough you can't see beyond the swill.

Angry? *Moi*? I don't think so (*exits left, tossing head and sniffing at a wholemeal nosegay*).

So, we planned to do our bit for Nature and the World, pretentious or otherwise.

Part of our plan was that we should have a patch of woodland. We've always loved trees, especially beeches. Would we strike lucky and buy somewhere with a stand of wonderful beeches?

Well, no, but we did have *some* trees ...

– 1 –
The Woodland

Woodland is a mixed blessing if you keep free-range poultry. Guess where lots of foxes live?

Our problem is really one of geography. The henhouse is about a hundred yards from the fence that divides our bottom field from the woods. But much closer is a patch of wild land, and the brambly ex-slurry pit.* There is a large hawthorn hedge and a patch of nettles next to the henhouse. If we had tigers in the woods, and once we did have wolves who'd escaped from a zoo, they could remain in perfect cover for all but five yards of the route from home to the take-away. For something as small as a fox, it's a doddle.

'Ah!' I hear you cry. 'Why not cut down all the cover?' Briefly, it would be a month's work, and still not be effective, because free-range poultry wanders about a lot. A fox can hide behind an apple tree. A lip-lickin' chicken wanders by and it's goodnight Vienna, hello Kentucky.

Fencing is likewise impossible if the hens are to be free-range. Anti-fox netting needs to be six feet high. We'd have to fence every yard of our perimeter which would cost thousands, and it still wouldn't work. One gate left open by a scrapman? And anyway, foxes are clever. They'd be sure to tunnel under it or paraglide over it.

---

* More on slurry later. Bet you can't wait.

No, there is no way we can have freelance poultry. The ultimate problem is the woodland, I gradually realised. Our land slopes with increasing steepness towards the streambed of the Siedi (a tributary of the Teifi, itself a tributary of the Irish Sea and Atlantic Ocean). There comes a point where it is too steep to be worked, and remains as wild woodland, which stretches for miles in either direction. Predators live in it, and plunder the farms on both sides.

It's very rough down there. Fallen elms, which we've not been able to get at to cut for firewood, treacherous boggy leaf mould, slippery rock, tangles of hazel, and whippy ash, densely packed between occasional mature trees, mainly oak. It's rather like a rainforest, in fact. Thousands of saplings striving for their place in the sun; and everywhere damp and humid. Thick pillows and wads of moss, strappy ferns, billows of bramble, and nettles over seven feet high.

It's a wildlife paradise. Nothing ever gets disturbed down there. We venture in every year or so, in March, when the undergrowth is still in hibernation; otherwise it's Nature's own. I've not heard Pan's pipes yet, but it's only a matter of time.

We still have plans for the woodland, but they're not pressing. We thought we'd coppice one acre for firewood and fencing, and thin out another acre to allow large trees to grow through, to provide building timber. The remaining acre we would leave completely wild.

Maybe next year?

Meanwhile, what of our two remaining ducks? They went on a foray last month. Up the lane and down inside another field. Couldn't find their way back through the fencing. We tracked them down by their feeble quacking and had to carry them home. They were too exhausted to walk.

Then they went on another foray, and all we found were feathers. Goodnight Bombay.

* * *

No beeches, then.

But we soon discovered a marvellous little grove on the banks of the Teifi, just three miles down the road, at Henllan Bridge. Here, the river narrows and spumes over rocks, and chafes out deep swirling pools, a beckoning challenge to the local canoe clubs, and to the occasional man (usually a man) overwhelmed by despair or grief.

The beeches form a still cathedral of peace, with soft carpet underfoot and Gothic vaulting above. At any time of the year they are profoundly beautiful. In spring the spiked buds unfold into downy green clusters; in summer the translucent leaves scatter shadows and a gentle rustle of tranquillity; in autumn the gold and russet foliage shimmers in a million tones; in winter the fine twigs interweave into a ceiling of tracery … and then the next spiked buds appear from nowhere. Just another miracle.

We visit the beeches a couple of times a year, just to *be.* York, Chartres, Henllan …

* * *

Our own woodland is different. It is clearly much more like the aboriginal forest that covered the entire land before Man hacked it down to feed his expanding and bawling family a couple of thousand years ago. And though it lacks the stunning atmosphere of the beech cathedral, it has a primeval presence all of its own.

It is an adventure to haul your way through the brambles at the rim and down through the spikes and potholes, scattered like man-traps, into the woodland proper and its heady fug of mould. There, you can hear the stream fifty feet below roar its way past after a storm, or clink and splash in more gentle times. It echoes off the banks and fills the valley. You can sit with your back up against a fallen tree and just … listen. It will not be long, methinks, before an open mind might perceive an elf or two, doing elvish things among the wild wood. 'Oy, stop that … Yes … you with the pointed ears and the silly hat. Just leave that hedgehog alone. Go on … *Go* on …'

It can be a bit more of an adventure than you were banking on if you go down the wrong part of the slope. One of the previous occupants has done what country folk have always done with their rubbish and chucked it down the cwm. That was fine in an age of oak buckets and rough pottery, when you were unwittingly piling up a treasure trove for future archaeologists, but it's an entirely inappropriate modus for the machine age. There's at least one old washing machine down there, which will take centuries to dissolve. There's also a derelict sofa or two. The fabric will quickly rot, after providing a deluxe high-rise for several generations of rats, but the steel springs will be a menace for decades. Most worryingly, there are long and hidden daggers of plate glass, just waiting to slice a wellie-rinded steak off your foot. You could lose hundreds of gallons of blood before you could haul yourself back into the yard. And then what? You'd lose hundreds more, scrabbling pitifully up the yard, and yet more as you drag yourself to the phone, *ruining* the lino, and then you find the ambulance is busy answering a hoax call somewhere in Aberdeen, and so you just *drain out* entirely, next to the fridge, and can't even reach up for the pad and pencil to write a farewell note.

So … I advise caution if you go down that way.

Our pre-teen daughter Cait used to go down to the stream a lot with Edward, the son of our neighbours and Guardian Angels, Ken and Doreen, who have been farming for generations, so to speak, and have forgotten more about sheep and cows than they could ever teach us.

Cait and Edward knew all the routes, and would wander for a mile in either direction, climbing fallen trees and wading through the stream bed when the overgrowth got too difficult. Cait says Edward showed her how to tickle trout down there.*

* Possibly by telling them jokes. She wasn't entirely explicit on this point.

I don't think there can be many trout in the stream these days. I guess there might be a few, but there's a sad expectation that chemical and biological pollution (mainly from slurry, of which, yes, more later) will have pretty well killed them off.

But years ago there were plenty of trout, and the preferred local technique for catching them was with … well, would you care to guess? An antique Hardy rod perhaps, with a selection of hand-tied hackles and fancy flies?

How about a sledgehammer? No, really. For a start you couldn't use a fly rod even if you could afford all that fancy gear, as there's absolutely no room to swing a line.

So what about the time-honoured grub on a pin on a string on a stick? Well, yes … it would probably work, but it's so much *effort.* Why not analyse the situation and think laterally instead, the way the locals did?

The task? To catch fish (not 'Commune with Nature'; you do that all day every day and need some time off. And not to look elegantly eccentric in a County sort of way in a silly tweed hat and PVC rompers; you have no time for such pretensions. And *certainly* not to outwit a fish. For Pete's sake …)

The circumstance? Narrow fast stream, with minor pools, obsessively overhung, with quite a lot of flat stones here and there.

The principle? Sound travels very quickly in water and can daze anything in it for just long enough to be grabbed.

The solution? Well, it's obvious, isn't it?

The technique? You stand quietly for a while, so your prey becomes accustomed to your presence; then slowly raise the hammer; then ***welt*** it into the middle of a nice big rock, just below the surface, like Thor cracking cockles. Any nearby trout will obligingly reveal themselves, and all but jump into your bag.

A word of caution: *don't aim for the fish*, as that way lies madness and disappointment.

If you wear particularly heavy studded boots you might manage without the hammer but would be certain to attract funny looks from any passing ramblers. My reaction would have been to tell them it was a traditional dance, and offer to demonstrate it again in six local variations, jetée-ing from pool to pool, arms akimbo, but the circumstance never arose.

I've heard of the sonic approach to fishing being applied elsewhere. One of my ex-college students was a Chinese genius called Peter, from Hong Kong. He was fond of hiking and camping at weekends (yes, in Hong Kong – apparently there are more wild bits than we imagine). He would always travel with nothing more than a knife (no tent even) and something useful in a waxed roll. Nothing else at all. He lived off the land, you see.

Being of a pragmatic nature, he didn't bother with making snares for rabbits or dragons, or whatever it is they have in Hong Kong. Fishing was much easier. But no fish hooks? 'No. Much better thing …'

In his waxed roll he carried a small bundle of home-made dynamite and a couple of matches.

'Very easy. You light little stick and … BADOOOM. Plenty fish.'

* * *

We could have made a considerable addition to the household

economy if we'd ever got round to carrying out the coppicing plan for the woodland.

Dad gave us a winch as a Christmas present for Fergie,* but the various pressures on us to cope with the day-to-day veg production meant we never quite had enough free time to find out how to work the winch and then give it a whirl.

The plan was to sit the tractor on the brow of the hill, run the winch cable down into the cwm, loop it round a suitably lopped elm trunk, and then slowly haul it up the slope to where we could butcher it in a welter of sap and flying splinters.

Students of physics will already have noted what fun the God of Action and Reaction could have here. We wanted the motor to haul wood *up*; clearly, there was an equal and opposite option of the motor hauling the tractor *down* – at an accelerating pace, until it bounced and hurtled into the cwm at forty miles an hour, obliterating any undergrowth, small mammals, or inept smallholders in its path.

Hippie Smashes Himself To Pulp With His Own Tractor ... In His Own Wood!!!

Photos Page 6, 7 and in Forthcoming Easter Supplement

But we were prepared for this one, actually. The tractor would be well dug-in, and chocked and wedged with at least fifteen different modes of resistance. It would have worked. Probably.

Always trying to think ahead, we investigated contraptions for converting a chainsaw into a crude planking device,

* Our little grey Ferguson tractor. A design classic par excellence (SfaS).

but decided against it. Too expensive, especially as we didn't yet have any trunks to plank. But it would have been great to have enough home-grown slabs and posts to build a mezzanine in the Big Black Barn. Then we could have the bales of hay and straw upstairs, and keep the downstairs for machinery, tools, broken washing machines, old sofas, plate glass daggers etc.

Coppicing would have been even more useful. A coppice is a stand of woodland, often of ash, in which the main trunk has been felled (and used for making a mezzanine, say) and the stump has been left to sprout again. The bigger the stump, the more sprouts and the stronger they grow. After six or seven years, a sturdy stump might give you four or five stems as thick as your arm, or a supermodel's thigh, which you can harvest for winter firewood. Just saw through, drag home, saw into lengths, and stack to mature for six months. No huge trunks to haul, no risky chain-sawing, and no even more risky splitting with an axe and chisel. And the twiggy bits make good kindling, or arrows, if you're thinking of holding a Crusade; and of course, the cow and sheep will enjoy nibbling at the leaves.

Obviously you need quite a few stumps to produce enough fuel for a whole winter, and six or seven times as many again if you are to have fuel for all future winters while the original stump regenerates, but we reckoned we'd be in with a chance if we had a full acre. At the very least, we could cut our coal bill in half.

Serendipity being what it is, what we have ended up with is three acres of wildlife reserve. That's OK. You can't win them all.

Pitching in

Cait was four when we moved in. Paddy was tennish. Anne (She Who Understands Things) was two years younger than me, and I was thirty-nine and a twelfth.

My parents were in their sixties and recently retired. They pitched in with us partly to be nearer the grandchildren, and partly to help with the cost of buying a property. Without them, we could never have afforded anything big enough to have realistic prospects. As it was, we paid £39,000 (well, this was 1982, before prices went ballistic) for our modest farmhouse, multiple outbuildings, five acres of green field, three of woodland, and a longish driveway, built along the bed of what used to be a stream. More on this gem of amateur engineering later.

We all rubbed along pretty well, but we were overcrowded from the start, and would sooner or later have to do something about it. In 1984 my parents, Ralph and Jessie ('Korki' and 'Poppet' to the kids), amazed us by getting

Agricultural Planning Permission to build a bungalow in our top field, and in the spring of '85 they moved into it. Now we were still close, but everyone had more room.

1985 was a busy year all round. Mum and Dad dived into sorting out their new home, and Dad began busying himself with his new bees. (My own impartial and balanced views on those sodding bees are made particularly clear elsewhere (SfaS) – more below actually, in Chapter 18, so you can see what I mean, and judge for yourself, also impartially.)

We at the Big House were learning about the complexities of running a house cow, keeping a few sheep and poultry amused, and earning enough money to pay our essential bills via growing and selling lots of vegetables. This may not sound like a big deal, but if you take a look at the price of spuds and calculate how many kilos you would need to pay your Council Tax, you'll soon realise that this is no trivial matter. Then consider the fact that the guy who actually *grows* the spuds gets only a fraction of the price they charge you in the supermarket,* and it soon becomes obvious that it takes a *lot* of veg to pay a 'real' bill. As a rough guide, a smallholder might need to plant, cultivate, weed, harvest and deliver 50kg of spuds (that's two big sacks-worth) to pay for one gallon of petrol. How many sacks would he need to grow to pay a phone bill of say, £75? How many hours' work would that represent?

That, of course, is one reason why we didn't grow spuds; not on a commercial scale, anyway. They are simply too cheap

* A supermarket paid an acquaintance 80p/kilo for his leeks, later put up for sale at £4.80/kilo: a sixfold profit margin.

per unit, because big farmers can use big machinery to plant and gather them. We use a fork.

The only way we could hope to pay our way was to grow crops that a mechanised farmer couldn't. Our original idea of garlic was near-perfect, in theory anyway. What a pity the summer of '85 had other plans for us.

So we tried courgettes instead: another crop that returns a reasonable unit price, and which needs gentle handwork, not a socking great £250,000 Harvest-o-Matic SuperTurbo that makes your supermarket spuds so ridiculously cheap, if you don't count the pollution from excess fertiliser, pesticides and fungicides, of course.

* * *

Apart from being busy with our new homestead, shunting furniture around, and cleaning up extra items that had been relaxing in a damp barn for three years, we pressed on with our grand plan for a self-sufficient closed organic system. We already had a cow, who provided lots of milk as a by-product of her main task, which was to produce lots of fertiliser. And we had a bevy of assorted sheep, mainly Jacob/Suffolk crosses to help keep the grass in fettle, as well as helping out with the fertilising, and eventually with the meat and wool supply. Also, we had a dozen or so poultry random and assorted, mainly for the eggs, but also for de-bugging the sward and veg patch. What next?

Well, we still had a lot of surplus milk, which we couldn't face chucking on the compost heap any longer, so we did the sensible thing and bought in a couple of…

– 2 –

Pigs

Are pigs worth it?

It depends what you want, I suppose. Some breeds are thoroughbreds: lean mean meat-machines with a barley-meal/pork conversion ratio that would be the despair of any ballet dancer, while others are amiable slobs, happy to gradually convert weeds and surplus biomass into a modest stack of bacon butties, while repaying all attention with a baritone grunt and a waft of aromatic hydrocarbons.

We raised a pair of piglets to porker age for three consecutive years and have mixed views on whether they are worth the time and trouble or not.

We optimistically transported the first pair home in the back of our Reliant Robin in a big cardboard box. They burst out of it in ten seconds flat and rocketed up and down the back of the van for the next twenty-five miles, squealing and piddling in all directions, and causing a couple of potentially dangerous swerves as they bounced and hammered into the back of my seat.

As soon as we got home, one escaped; see below …

Once impounded in their paddock, they decided that their first job was to ricochet around inside their sturdy wooden hut* like rock stars in a Wendy house, until it burst

---

* An old packing case for a big woodstove, which had previously done sterling service as an indestructible chicken bunker.

apart. They then zeroed in on the pigmesh fence and began systematically rootling their way under it. I was reminded of that scene from *The Great Escape*, with myself cast as the Snooping Goon, while Lardy and Stinker act as lookouts for hordes of desperate pigs unseen, all busily tunnelling under the mesh and into the potato patch a mere thirty yards away.

But we thought we'd try again and determined to outwit them the following year.

I patched up their hovel as best I could, and installed some primitive electric fencing just inside the pigmesh. This consisted of 18-inch lengths of batten hammered in every ten feet or so, with loops of bike innertube tacked on top as insulators. Through the loops ran the standard steel-threaded polypropylene electrowire run off a battery-powered fencing unit we found in a hedge. The pigs, 'Piggy' and 'Piggier', walked straight through it, without so much as a whiff of BLT.

Sterner stuff was required: so we bought a little mains-powered unit and ran a line 150 yards across the field from the ticking yellow box on the kitchen wall, and bought a reel of expensive multicore *Ol' Sparky* professional grade electric fencing wire. I redesigned the battens and toshed them in at 45 degrees, with the tripwire glinting just six inches off the ground, then hobbled back to the cellar, wild-eyed and cackling, and threw the mighty switch. Within seconds an outraged screech told us all would henceforth be well, and that we would be spared the tiresome prospect of combing the railway station at dawn to check for two porkers on the run in dodgy moustaches and trilbies.

As for turfing out grass and weeds, fertilising as they went ... well, they didn't. Instead, they dug several foot-deep trenches, while leaving everywhere else completely untouched. They also compacted the soil amazingly, especially after rain, and insisted on dunging in only one remote corner. Clean, you see. Not 'dirty pigs' at all. Thus hopeless at muck-spreading.

Oh … what mayhem they caused to the land: pits and wallows surrounded by dimpled concrete that the rotavator skidded off, like a plastic fork off a bouncer's forehead. I once calculated on the back of an envelope that the down-force exerted per grubby little trotter would be in the order of fifty pounds per square inch. I seriously think a Volvo parked on your foot wouldn't do as well.

It became obvious that we'd never be able to reinforce their old hut well enough, so we kept the last pair in an old open-fronted shed, bedded on straw and torn up *Daily Telegraphs* donated by Dad. They were very happy, and thrived, and liked playing with their plastic football until one day it disappeared. We didn't like to ask.

All the piggies went for pork, and we were truly grateful. Wonderful meat, and Anne made a chunk of the best bacon the world has ever seen; but hard work and not much cheaper than the shops, except that the quality really was incomparable.

And then there was all that expensive barley-meal … surplus milk … starving children …

Ah! The One That Got Away: naturally enough it was raining when we finally rocked to a halt with two loose cannons barging about in the back. We penned one quite easily, but the other slithered away and instantly scuttled off up the drive to Freedom. Have you ever tried catching a wet piglet? There's nothing to get hold of unless you get down on your knees and grab an ankle. And by the time you've got to your knees, the ankle is already several feet away, and accelerating. Off it went, tritty trotty. I followed, but by definition could only chase it further and further away. Then it dodged through a tiny gap by a gate into the field full of Jersey heifers.

If we'd filmed it we'd have made a fortune. Picture, if you will, a tired man, with a bruised back, clad in a knee-length coat, home-waterproofed and thus amusingly mottled (for details on how to make your own see Chapter 14); wellies on feet, one green, one black (ditto Chapter 27); droopy bush-hat on head, wringing wet and ridiculous; glasses streaming with water and smeared with mud. He is stumbling as fast as he can in pursuit of a small and cocky pink piglet, who is effortlessly keeping ten feet in front of him. They are crossing a twenty-acre field, uphill, towards the Dingly Dell and a chest-high barbed wire fence. Six feet behind the man is a herd of fifteen frisky heifers, fighting for the yellow jersey.

Exhausted, the man leans on the fence, dripping from every hem and protuberance, surrounded by doe-eyed and jostling cows, while the pig squeezes under the fence, tumbles down the six foot drop, and trit-trots off down the lane, whistling…

The man has just enough energy and patience left to know for a fact that hauling himself over a barbed wire fence, six feet above a tarmac surface, would be a very foolish thing to do, particularly as every item of his clothing is wet and bagging and absolutely certain to become snagged at least twice. He could, he knows, be hanging on the old barbed wire for days before anyone found him.

But who is this? The hippies from the cottage in the

Dingly Dell! 'Stop that pig!' One reaches down and [illegible] does so. Just like that.

I should point out that the cloudburst had now ceased, and the hippies were all perfectly dry and full of summer fun. How nice.

Anne, meanwhile, roars up in the Reliant, and slews it to a diagonal halt in the lane, Starsky and Hutch style. 'OK Schweinchen … for you der Krieg is over!' A noble effort, but quite inappropriate as a fair-sized horse could have by-passed the Reliant on either side, thanks to the generous ditches. But the thought was there.

Then we all drive home, pig and wife grinning, man dripping and fuming.

Some weeks later I visited the hippies, with a very tasty ham. After all, they had saved our bacon.

* * *

Actually, come to think of it, we did try making the second pair of porkers a Grand Design of a bivvy out of a dozen nice warm straw bales topped with three or four absolutely waterproof curved and corrugated asbestos-cement roofing panels. Airy … warm … spacious, even. (The asbestos is perfectly safe, incidentally: it's locked into the cement, and can only escape if you gnaw it with your teeth, and even pigs can't be bothered to do that.)

A nice warm den then?

They barged it apart as soon as they looked at it. Bales chucked about like dominoes.

So then we moved them into a brick-built shed. Open plan, draughty, possibly even a trifle damp in one corner …

heavens. Did they notice? Nah. But at least they couldn't just barge it apart.

Daisy, Daisy ...

We'd had Daisy for a couple of years by now, and had learned how to hand milk pretty well (SfaS; plus Chapter 25). We'd also learned what hard work it can be, turfing out a whole winter's worth of bedding when heavily steeped in pee and slurried dung, and then trampled into a thick moist cake. More on this later too.

Daisy was knocking on a bit, though. She'd already had half a dozen calves before we bought her, so we knew we would need to replace her eventually, preferably with one of her own daughters. Her first calf with us had been a little Jersey bull, called Wally, who would obviously not grow up to be a milker. We hoped her second effort would be more ... helpful.

It turned out fine. On April 1st she gave birth to another fabulous little Bambi of a calf, just as beautiful as Wally, and much more valuable to us. Cait and Paddy witnessed the whole astonishing process. A useful life lesson we hoped.

I wanted to call the newcomer Tallulah, but Anne had already decided on calling her 'April'. OK, April then. Boring. Boring. Could have had Aimless, Pointless, Graceless or Feckless, as a homage to *Cold Comfort Farm* ... but April it was. And it wasn't a boring name at all. It was light and airy and lovely, and it matched our newborn a treat.

She was a delight from the start. Lambs are cute, piglets are entertaining, chicks are charming ... Jersey calves are heart-melting.

April's job for the summer was to drink as much of her mum's milk as possible and to grow strong. Daisy's job was to protect her, and most of the time, she knew it.

There was one occasion though where she simply wasn't concentrating. Towards evening, after a hard day's grazing, Daisy realised that her udder was beginning to need relief and looked round for April, who was usually more or less at heel, possibly doing a little light pronking and dashing round in erratic circles. But there was no sign of her.

We realised what was going on when we heard Daisy's plaintive little moo. She looked all around, and gently mooed again. What we knew, but she didn't, was that April was fast asleep like a fawn in a forest glade in some longish grass just ten yards away.

'Oh, Daisy,' I said, 'what have you done with your baby?'

She glanced at me then looked away again and mooed. I'll swear she was embarrassed.

'Where is she, Daisy?'

Moooo.

'You've lost her, haven't you?' She looked straight at me. I'm sure I detected an anxious request for help in her eyes. She was just this side of distraught.

'She's over there, Daisy,' I said, pointing to the long grass. This was several steps too far for a cow, however. Following a pointed finger, or complex directions, was way beyond her.

'Come on then, follow me.' And she did. This was surprising. After all the lip she'd given me, and on one occasion, spilt ... she was finally bowing to my superior wisdom.

April was still spark out, despite all the anxious mooing. Mum nuzzled her and woke her. Whether she then apologised for losing her I don't know, but she seemed to keep a closer eye on her afterwards.

As for defending her calf, I was once astonished at Daisy's instincts. She was a few feet away from the dozing April, peacefully nibbling.

I happened to be passing and wondered what Daisy would make of a skulking figure. So I paused about twenty feet away, then hunched forward. Then I took one slow and cautious step towards her. Then another. She could see me out of the corner of her eye.

After my third stealthy step, still hunched forward, she raised her head, took a half step forwards, tilted her horns, and bawled at me. It was quite alarming, and served me right, of course. But I'd got an answer to my question.

Even though she'd known me for a couple of years as a friend and guardian – who had even once helped her find the calf she had so callously abandoned – still she saw me as a threat just by virtue of my body language. How? This was a domestic beast, born of a long line of domestic beasts who had never seen anything more threatening than a squirrel, or possibly a sheepdog, in her long life. She'd never ever seen a predator. Yet she still could recognise shifty and potentially aggressive posture. How? 'Instinct'? But 'instinct' is just a word. It doesn't even begin to explain how a cow removed by a hundred generations from the time of wolves and bears could nevertheless recognise 'threat'. DNA is no answer either, as nobody has a clue, even in principle, as to how a lifeless chemical might store a memory of any sort, let alone what amounts to an ancient folk memory, latent for thousands of years.

– 3 –
The Rockery

I'm not very good at flowers. It's not for want of trying; it's just that the names don't seem to stick. 'Cabbage'? Got it. 'Rhubarb'? No problem. But *Spiridophignothium shufflebothamii*? Sorry, it's gone already.

It's largely to do with those long Linnaean names, that all sound like minor Roman emperors: *Panspermia rampans*, and so forth. I cope very well with Daisy, Dandelion and Buttercup, however, and even with some of the more obscure floriana like Stitchwort and Hawkspit.*

I've gradually become used to my inability to remember flowers' names and regard it now as just another in a long list of minor personal failings, like not being able to mend cars, fill in forms, or cook.

But I do like flowers, and one of the great delights of the summer this year was the rockery Anne made in what we used to call the yard, but obviously now call The Patio. She made it by creatively collapsing part of the retaining wall which keeps the lawn four feet above the yard. As this procedure was clearly not going to grow any more tasty vegetables,

---

* I've just been informed that that should be 'Hawksbit', a version no more meaningful than 'Hawkspit', in my view, and considerably less romantic. Anyway, which 'bit' of a hawk are we referring to here? The round yellow fluffy bit that bees like, and which shrivels and drops off in the autumn? Or some other bit?

I originally wrote it off as a mixture of mindless vandalism and a simple waste of time. 'I don't know what you think you're doing. You're only encouraging entropy, you know,' was my clinching argument. 'But it will look nice,' Anne countered, rather wetly I thought. 'Pshaw,' I said, and 'tsk!' and swept out.

Needless to say, she was absolutely right. A boring wall has been transformed into a cascade of shape and colour, stuffed with plants she's bought, cadged, or stolen from friends and ex-friends. There are beetles and bugs, dozens of wild bees, waspy things, and the occasional lizard. There are wrens nipping in and out, and, as the Cranesbill ripens, bullfinches eating the seeds; and a couple of sleek, streamlined, fawn and beige jobs which might be warblers. There is also a small painted concrete duck, which is my own aesthetic contribution.

My other contribution was a couple of plastic chairs (in 'Blinding White', to match the pottery gnome, whose paint has all washed off) which are the perfect complement to the rockery. Before, it was *there* but you couldn't really see it without making a particular effort to go outdoors and stare at it, as a purposeful act, like mowing the lawn, or blessing the compost. But now you just plod out of the door with your tea and flop into a chair; and spread out before you is Anne's inspired waterfall of colour and pattern.

I'm still not very sure what all the plants are. Pinks I know, and *Allium*, and planted between them, inter *Allia*, as it were, those long tall, rather elegant things with purple flowers: *Aquilegia*? Then there are Poppies, a Foxglove or two, Geraniums, Teddy Bear's Ears, and Blue Tufty Jobs. That's my lot. But it can be fun when we have visitors. 'And what's that one?' 'Oh, you mean the *Alopecia*?' I reply airily, 'next to that sweeping cascade of *Streptococcus* and the miniature *Drosophila melanogaster*? Very rare, that one.' Most men are impressed by this deep knowledge of botany; but not so the women, who either raise an eyebrow, or more often smile quietly while hubby pokes at the *Gingivitis flatulans* with his air-cushioned trainer.

When it comes to flowers, as with many other things, women know their *Allia* … and how to create beauty out of nothing.

* * *

Yes, I still think women are a jolly good thing. *Ow!*

The rockery has gone from strength to strength over the years. Anne's first principle of gardening is *Let It Get On With It*. This means that plants get stuffed in where common sense suggests they'll thrive, and if they don't, they don't. If they're obviously failing, they are likely to be moved to a new site to try their luck again. (No such thing as a site for life these days, anyway.)

Her second rule is *If It Spreads … Good*. A rampant plant cuts down on the weeding, and if it gets too invasive, you can always yank out a few yards and that's that: weeding done for the year.

Rule 3 is *There's No Such Thing As A Weed*. Thus we have Daisies and Buttercups among the Blue Tufties and the *Crocosmia*, and the occasional Guest Plant like Sweet Cicely or a deeply sinister-looking item with a dull purple flower and dull green leaves. I can't remember its name, but I bet it's

Whorebane or The Leveller, or some such moniker left over from the Middle Ages when plants had uses above and beyond mere food. It oozes bad vibes.

The overall effect of the rockery in the winter is of natural sprawl. Dead ferns provide cover for the first spears of the daffs, and the *Alchemilla* looks as if it's hibernating: brown and drooped. No doubt we'll tidy up when the new plants really need a bit of help. Meanwhile they shelter bugs and beasties who are very welcome to share the garden, as well as being food for the wrens and finches.

The thrushes appreciate the snails too. Not so the slugs. I once saw a thrush, a youngster I think, driving itself barmy trying to unglue a slug from its person. It wiped; it swiped; it smeared; it poked and pulled. All it did was to shift the eviscerated corpse back and forth round its beak. In the end it flew off and presumably offered the snack to one of its friends. Then made itself scarce, sharpish …

My final contribution to the decorative effect of the rockery is a Tableau of Europe, made up from occasional artefacts picked up at boot sales. Centre stage is a six-inch Eiffel Tower. To its left is a chubby Germanic boy in lederhosen and a hat with a flower in it. He is clearly eyeing up the Tower. I leave the symbolism to you. On his head is the tin top off a bottle of Cava, carefully wired on. It *looks* like a steel helmet, but in fact it is a symbol of economic power …

To the right (yes, 'the right') of the Tower, and a foot or two distant, is a two inch model of extraordinary design: a silver bucket, containing a rosy-cheeked baby in a frilly mob-cap. The bucket is filled with 'water' (yes, *water*).

The baby's expression is cautious and apprehensive, and also resembles the late Winston Churchill more than somewhat.

Put all this together, and I think you'll see what I'm getting at.

A couple of feet to the left of the boy are a pair of white and yellow cement ducks 'floating' on the stonework, and looking on with intense interest. They too are wearing Cava helmets (remember the 'economic' symbolism?) To me they represent the Japanese Navy.

So there you have it. Something for the Eye and the Soul, and something for the Mind as well. We work well as a team, I think.

Best of all we have a foot-high Chinese Buddha, who sits cheerfully and disengaged come hail, rain and snow. A variegated Ivy climbs gently onto his lap, but never seems to overwhelm him. A Periwinkle stars him with blue from time to time.

ME and my health

The big issue for Anne and me in 1986 was my health. After the pole-axing just before Christmas, we needed to know whether I'd be up and running again in time for the madhouse of springtime sowing, ploughing, planting, weeding, and all-round five-star panic, which traditionally began with trying to get the tractor started without the help of sorcery.

The good news was that if the Lords of Karma decreed that I just *had* to get ill, because it was good for me and I'd thank them for it in the end, then Christmas was just about the best time possible for it. I'd have three full months to recover in.*

The bad news was that this wasn't a common cold or a dose of headbanging flu. ME was a condition nobody knew anything about, so they'd done the only helpful thing possible and given it an unpronounceable name and hoped for the best. The symptoms reported are so various that it scarcely merits a name at all apart from 'General Overwhelming Knackeredness', or GOK. Some people have awful pains, and just about every sort of unpleasant symptom you can imagine. All I had was a total loss of energies, endless turbulent dreams, and an acetic aroma which even the dog turned her nose up at.

Obviously, my physical energy had fallen through the floor, and because I was so sleepy and gormless, so had my mental capacity.

And yes, I did feel just a tad sorry for myself. I was even more sorry for Anne, as she not only had the worry of her husband being quite alarmingly ill with something that was just a name and for which there was no treatment, but also had to look after the kids and the home, and all the animals, and had the prospect of doing all the imminent spring-work on her own. Just for starters, this meant making up fifty kilos

* That's the way farming works. You don't actually have four seasons but only two: waxing (yangtime) and waning (yintime): or, if you prefer, spring-summer, when everything goes ballistic, then autumn-winter when everything needs collecting and sorting. 'Summer' per se is for endless coping and TLC, and 'winter' is when you go and hide and catch up on all those weekends and bank holidays you couldn't afford to take off throughout the rest of the year. And mend sheds. And dig ditches out. And fell trees and chop firewood for next winter. And ...

of seed compost from peat and additives, then filling 3,000 or so three-inch modules with it, and then sowing a courgette seed into each one. Then all these trays needed ferrying into the polytunnel and carefully watering. That would be the main crop under way. Next came the sowing and planting of every other crop: dozens or hundreds of tomatoes, lettuce, potatoes, leeks, and many others, which meant land preparation as well as all the actual seed-work. She'd previously (and very sensibly) left the ploughing to me, and as now was not the time to teach herself, her only option was rotavating.

Rotavating isn't actually difficult, but you do need to have a certain amount of brute strength to cope when the machine hits a cat-sized rock just below the surface, or even a particularly thick clump of coarse grass. It can easily tip over if you don't forcefully guide it round the problem and if you're not sharp on your toes, it can pull you over on top of it when it goes (see Chapter 9). It can then helpfully catch fire if the fuel escapes in the wrong direction.

But she did it. The first thing I knew about it was when she came up to the bedroom with another cup of tea and said, 'The rotavator's stopped. I've checked the fuel. It's not the fuel. Sorry.' She was also rubbing her wrists a lot. Shoving a heavy rotavator round a stony field is no work for someone with no great body bulk and incipient arthritis.

I got out of bed and had a look. Seized. Although the Howard is a sturdy machine, it was using oil more than it should have and as the sump held only a trifling amount it could quickly run completely dry and seize up if you didn't keep a very sharp eye on it. In her hurry to get another job done and crossed off her endless list, Anne had forgotten to check it.

Well, while I was up I might as well do something useful, mightn't I? I reversed the Reliant up to the bottom of the field and shooed a curious Daisy off.

The problem now was how to get a seized-up machine out of the cultivated area and across the meadowland, then into the back of the car. 'Into the car?' Yes … because there was no chance of me trying to fix it myself. Even if I'd had a rudimentary idea of the theory (which I didn't), I was too zonked and gormless to know which hammer to select, let alone which nut to apply it to. Alun was my man, and boy, was I grateful that he was.

Fortunately it was downhill all the way from the muddy place the Howard had given up in, and at a sloping angle across the sward to the car. Then there was a problem, as the floor of the car was a foot or more off the ground and the rotavator weighed a metaphorical ton: more than we could lift between us, at any rate. And because the engine wouldn't start, we couldn't just drive it up a couple of boards and into the vehicle, like a tank onto a landing craft.

We could just about *push* it, though.

Anne fetched a couple of our mini-scaffolding boards. They were about five feet long and plenty wide enough, but – as any aspiring trigonometrist will have already calculated – a five-foot long hippopoteneuse is a mighty steep slope to raise 100+kg up by a foot or more. Sisyphus would have baulked. We didn't. We had no option. That land had to be rotavated, so the rotavator had to be fixed. Alun said he could do it straight away if I could get it to him, so … get on with it …

I leaned on the handles and pushed, while Anne grabbed hold of the casing and shoved. It moved an inch or two, mainly sideways, as Anne couldn't get her weight properly behind it. Then I clambered into the car and heaved while Anne did the pushing. Ah! Promising, but useless, as the machine was not yet far enough up the slope for me to get a proper purchase on it. And the planks were beginning to wander out of line.*

Plan C: I leaned and pushed the handles while Anne leaned on my backside and pushed me.

Cows are very curious creatures, and Daisy was particularly so. Halfway through the Big Push Anne suddenly yelped and swiped behind her. Daisy retracted her horn with a theatrical flourish and, I swear, chuckled. 'Garn; *Gertcha*!' Rude of me, I know, but effective.

This time, definite progress. We shunted the rotavator a good halfway up the slope, and I leaped slowly into the car, just in time to see the machine roll back down to the ground, with Anne struggling personfully to stop it.

Cup of tea.

Next time we got it right, with the aid of a brick to chock the machine halfway up the slope. Once in the car I could haul while Anne pushed. In, but only just. It felt horribly as though the car was going to tip up backwards and end up on its behind, begging like a puppy.

But it didn't. The fibreglass floor creaked a bit, but it held

* Speaking of Reliants and planks, it is apparently a fact that more Reliants are written off nose-diving into inspection pits than on the road. (For the benefit of anyone who's never had the joy of owning a Reliant, it has only one wheel at the front: a fact that is easily forgotten ...)

the weight and that was that. In with the boards and off to Alun's garage.

Obviously I had to put up with the customary bit of mechanic's banter: 'Still got this old plastic thing? Where's my lighter …?' but it was a cheap price to pay. Alun dropped the job he was doing and helped me haul the Howard onto the yard.

'Briggs & Stratton, is it? Tough little motors. Should be alright. Here's hoping, eh?' and he began removing lumps off it.

It's a never-ending source of wonder to me that whenever I watch Alun fixing something, I can see what he does and he kindly explains what he's doing and why, but whenever I'm faced with the same problem again, I can never remember what to do. Never. It's not good for one's self-confidence.

Anyway … Alun spannered away and heaved and tugged at oily things, pausing occasionally to examine a particular piece like a zoologist discovering the previously unsuspected brown-kneed variety of the *Clatteroptopus* beetle.

'That's the spleen, isn't it?' I suggested.

'No.'

'The con-shaft?'

'No.'

'Well give us a clue.'

'It's the piston.'

'Piston. Got it. A round hole with a dirty flat bit at the bottom. Got it.'

'That's the bit that's seized. Oil should go round the edges to make it slippery and when it doesn't, it stops going up and down like a **** in a ****** and it stops and jams solid.'

* Technical terms which I won't bore you with.

'Jams. I see. "Seized".'

'Seized. You've got it.'

Then he fetched a hammer and an oil can and advised me to look away.

A few minutes later all was well. He reinserted the con-shaft thing and spannered all the screws up tighter than a duck's ****.* A couple of tugs on the starter string and the Howard throbbed back into life. I felt as relieved and grateful as any workman who has had his prime tool returned to him.

I paid Alun the trifle he asked for (a very generous as well as able man) and he helped me haul the machine back into the Reliant. He banged on the roof and I was off.

He called after me: 'Could have been worse …!'

'How do you mean?'

'… could have been mine!'

* See previous footnote.

– 4 –
Dipping

Until quite recently sheep dipping was compulsory. Twice a year every woolly in the land was forced into the indignity of being submerged in a filthy-smelling slew of dilute organo-phosphates (OPs), chemicals which became developed as nerve toxins. (Let's be clear here: a nerve toxin is a Weapon of Mass Destruction intended to *kill* lots of *people*.) This ghastly baptism meant that the sheep would be protected against scab and flystrike, conditions that lead to foul sores, debilitation and frequently to death; and we, the public, would thus be protected against, well … I'm not quite sure what. Debilitated chops? But unfortunately, those organo-phosphates began to be accused of causing debilitation in farmers. Not surprising really, given their pedigree.

Eventually the Ministry decided that the rate of scab had fallen enough, and the rate of unease about OPs had risen so much, that routine dipping could be stopped.

Bad news for the Aggro-Chemical* industry of course, as a thick mulch of paranoia means lots and lots of profits. But for once, reason triumphed over fear and OPs more or less bowed out.

---

* Yes, with two 'g's. Very powerful and very pushy companies these, who fund a lot of agricultural research. The results of this research are, naturally, entirely impartial. Oddly though, none of it ever finds in favour of organic as opposed to chemical farming.

Just before the compulsory dips were abolished, the organic lobby made a case for using some sort of non-OP solution based on lime, I think it was. Had OPs always been strictly essential, then? Or had they been used for so long just out of habit? And had anyone from the Ministry seriously looked into the lime option? Or had impartial research (see footnote) pooh-poohed the idea of a non-OP solution? No idea.

Did dipping work? Yes it did, and it still would, I guess. But these days the normal cure is an injection into the particular sheep, and not a blanket immersion for them all.

I know I speak for thousands when I say I don't miss dipping. The routine across the nation was for proper sheep farmers to dip their own wards first, then open their facilities and depleted chemicals to smallholders and Extreme Swimming enthusiasts.

This sharing process was necessary because a dip-bath is quite an expensive piece of kit that only proper farmers could afford to build. It's basically a 6ft by 3ft deeply sunken concrete bath, finished vertically at one end, and gently ramping upwards at the other. The victims ('clients', these days, I guess) queue up at the vertical end, and with a grace and artistry wonderful to behold, are gently but firmly upended into the bath by the farmer, one at a time, with marks being awarded for complexity of dive, loudness of bleating and bellyaching, and sheer bloody-mindedness in refusing the compulsory move of 'ducking the head completely'. A placid sheep would score only two or three points; a mover and shaker might chalk up fifteen or more. I believe the UK record is held by a Suffolk-Texel cross from somewhere in Gwent who chalked up a whopping 46, give or take, year after year.

Upon her demise, she was forcibly stuffed, and now serves as a footstool for eight of the regulars of a local pub.

Our dipping venue was with Ken, just a field away. Not too far to go, but …

First cross your field, with half a dozen nervous sheep, led by the redoubtable Cheeky, the doyenne of escape-and-scarper artistes. How would you approach this problem, dear reader? Train a dog? Dig a tunnel?

What we did was to make each sheep a collar and lead of baler twine, then bodily haul, guide, poke and gently belabour them across the two hundred yards of grassland. We never lost a sheep, but it took five of us to get them over there, and if one sheep decided to just stop (Bramble liked doing this), and splayed all four feet out, like a furry coffee table, it could develop into something of a pantomime, with five people pushing and heaving, while their charges wandered in and out and round and through, getting their leads impossibly tangled. Nightmare on Maypole Street.

Eventually they were funnelled into the holding pen, where they huddled in a corner and practised looking cool while voiding a lot of bodily contents.

Then Ken took the tin sheets off the font and we were off. Anne guided them forwards, Ken pitched them in, and I ducked them. For this job I wore as many wellies as I could manage, a long thick mac, a big hat, and some old gloves. My wand of office was a long crooked iron rod with which I had to force the critter's head completely under, just for a moment. It could then haul itself slowly and wringing-wettedly up the ramp and out to the draining and dripping pen, looking shocked and, frankly, resentful and abused.

Then it was a cup of tea and biccies all round (no biccies for the sheep), and we began the long haul back home. Invariably, there was a lot less whingeing this time, apart from the occasional 'Are we nearly there yet?', and some rather damp grunting.

Compulsory dipping has now gone, but it's not quite forgotten … because a number of people think organo-phosphates might be connected with the rise in ME. I always got splashed while dunking with the crook; and I know my gloves weren't always waterproof. And I did get ME.

Was there a connection? Maybe 'yes', maybe 'no'. Lucky dip time.

* * *

The Aggro-Chemical industry thrives on a farmer's fear, but much of this fear originates in the mind of the farmer from other sources. How did this come to be?

Put yourself in the place of farmer Jim Herrick, for a moment. His family has been on the same farm for several generations. Back in the 1850s, Jedediah Hayricke lashed out on a new means of cultivation – perhaps a deeper plough and more horses, and discovered that he could make good money growing a new strain of wheat and selling direct to a big city miller, cutting out the middle-man en route. He made enough money off his twenty acres to buy another ten off a neighbour. And over the next few years, through a mixture of good practice and hard work, and a bit of commercial savvy and enterprise, he built his farm up to a hundred acres: quite a substantial holding in those days. His farm was essentially like anybody else's: a mixed farm of livestock and cereals and vegetables and fodder crops. But it was the speciality wheat that made him the money.

Even though he was tempted, Jedediah knew that a hundred acres of wheat would be too much to sell or even to handle easily, so if he wanted to keep the guineas rolling in he needed to branch out or specialise in a couple more directions too. Recently the local town had developed a foundry and had just welcomed the new-fangled railway in, which was bound to mean an influx of people and an easing of the trade routes throughout the country. Not only would this give Jedediah a larger and more varied market to sell to, but all those new people would need feeding. Jedediah went in for some serious vegetable cropping, and started up a market garden. The veg would supply not just the local town, but many towns around, via the railway.

Jedediah's milk, too, could be railroaded into the cities. So could his livestock, direct to the butchers. Then he did the value-added thing and employed his own butcher. Jedediah got rich.

He was a hundred per cent organic, of course, because chemical fertilisers were still in their infancy, and anyway, Jedediah knew what every farmer knew: that animal muck is the best fertiliser possible. All his veg trimmings and surpluses went into his cows and sheep, so nothing got wasted. He employed his own pigman too, and several other farm workers, skilled, semi-skilled, and casual (at harvest time).

Eventually he set up a little brewery. Hayricke Hall became a minor industry.

He passed it all on to his eldest son, also Jedediah ('a man has the right to be proud of his name') who farmed in the same way as his father, and gradually moved with the times, cautiously feeling his way so as not to fall foul of fads. And always the top notch wheat paid all the bills and wages.

His own eldest son, Enrico (a story there, somewhere ...) wasn't really interested in farming, and took what to his father seemed an excessive interest in the theatre. He signed up in 1915 and was squandered on the Somme. The farm thus fell to Joseph, the second son, who should by rights have gone into the army or the church, or even the BBC (which had not yet been established, however). Joseph also wasn't interested, and the farm languished. However, there was a great demand for wheat for export, so Joseph bought into the new technology, sold off the remaining livestock, sacked as many staff as he could, engaged a professional farm manager, and went for broke on wheat. He thrived and even bought a hundred acres more.

As mechanical machinery and agricultural science developed, the family continued to thrive. Huge machines cropped huge quantities, and the magical chemical fertilisers kept the fields producing. True, the soil was getting harder to plough, as the organic matter disappeared, and was reverting to the clay that old Jedediah had turned into crumbly loam a century ago, and true, young Joseph needed to apply more and more fertiliser each year, just to maintain the same cropping level. True, too, that pests and diseases seemed to come from nowhere and multiplied extravagantly. But the new chemical controls were wonderful at keeping the pests at bay. Science was the way!

Young Jeremy, Joseph's firstborn, held the same philosophy, but even so, he wasn't too happy with his future. He knew his productivity was dropping little by little, and was only sustained by increasingly expensive dressings and dowsings with chemicals. He knew his soil quality was disappearing. He also knew he was dangerously dependent on the market for just the one crop. What if the bottom dropped out of it? Modern massive machinery, plentiful oil, and huge bulk carriers meant that Canadian wheat was becoming dirt cheap. What could he diversify into? Everything he looked at seemed ominous, largely because the Big Boys at home and abroad could always produce more crop quicker and cheaper than he could: Argentinian beef, for example. And global politics seemed to be playing a larger and larger role in the world of food. He passed the farm and his worries on to James, his only child.

Jim was on the ball.

Supermarkets would be the lifeline! Bulk buyers and bulk sellers! But they didn't want wheat. What then? Spuds? Yes, they'd want spuds; cabbage; onions. So the lifeline was there, but there was a snag.

He knew nothing about growing vegetables. All his capital was tied up in wheat machinery, and who would want to buy it? All his neighbours were facing the same problems he was.

All he knew about spuds was that you needed machinery to plough (which he'd abandoned in the era of direct-drilling, whereby the corn seed is rammed straight into the hard earth); handling equipment for the tons of seed potatoes; suitable barns, quite different from his grain silo; equipment to plant the spuds, which would need at least two hired hands as well; spraying booms and fertiliser spinners (ah! already got them, at least); special gear for lifting the crop out and then getting it off the field and onto big new cages – which would need human help as well; then handling gear to shift the spuds out of the cages; storage facilities for hundreds of tons of crop, which would need to be vermin-proof and dry; then machinery for weighing and bagging the spuds; and a bigger lorry for shifting them.

And then there was the problem of what gear you would need for cabbage; and then onions. They all need different equipment, facilities, and handling.

He took the plunge and started with potatoes. The supermarkets loved them and bought the lot. But he knew he was vulnerable, relying on just the one crop. So he saved hard and bought some expensive cabbage equipment. Now he could breathe a little easier.

The following year, the supermarkets told him his spuds were too expensive, as they could get them cheaper from the next county (where they had set up a dozen other farmers, just like Jim, and then played them off against each other), or Ireland, or Peru. He would either have to drop his prices or increase his productivity. The friendly rep from the Aggro-Chemical company assured him that a third dressing of *x* and an extra couple of sprays with *y* and *z* (the new super-pesticide, which required an expensive protective suit) would be just a little more expensive, but his productivity would rocket. Did it? Well, no. It did creep up a little, but the chemicals were getting dearer every year, and his soil was getting weaker. He realised that his land was essentially drug-addicted. If he cut the chemicals, his crop would plummet.

Then the supermarkets announced that they would not be buying his twenty tons of cabbage because the variety was 'too large for the taste of the modern housewife'. He should come and collect them from where he'd delivered them to. A dead loss, and an appalling shock; and another hundred tons of unsaleable cabbage still on the field.

The following week, they sent him a letter telling him that customers were demanding that their spuds should be no larger than 10cm and no smaller than 7cm, *and* washed, and then packed in little plastic bags. The 'grading' would mean that half his crop would be too big or too small to sell, and it would cost thousands to re-jig for the washing and packing gear. He took out a loan …

Every year, the pattern was repeated. He was now at the mercy of the supermarkets, and he knew it. Many of his neighbours had given up the battle and sold out to corporate agri-business. Jim wouldn't sell out. Old Jedediah was whispering in his ear.

He battled on, until he realised that his projected income for the following year could never exceed his bills, no matter how he fiddled the figures.

And that is the story behind why one spring morning Jim Herrick was found in Jedediah's little ash copse which Jim had resisted cutting down, despite the insistent advice of the Aggro-Chemical rep, with half his head blown off.

* * *

So it's all the Aggro-Chemicals' fault is it? Well, no; not really. Nor is it entirely the supermarkets' fault. They are, after all, just doing what capitalism expects of them: competing and cutting costs, and making the biggest profits possible. They are only, as they say, giving the public what it wants. And even allowing for the brainwashing we get from advertising, they are broadly speaking the truth here, because the fact is that we, the public, don't care about Jim and his like. All we care about is getting the cheapest possible food, no matter how it is produced, or what it's tainted with, so we can spend the rest of our money on essentials like £600 handbags and ten new CDs a week.

I've actually watched a woman in a market poke and prod at three lettuces, priced at 15p, 18p and 21p. She finally picked the 18p one, then turned on her heel and plucked a bunch of flowers out of a display vase for £1 without a second thought.

As long as we have so little respect for the quality of our food, we will be served the rubbish we deserve: the Turkey Twizzler Syndrome. Too bad that our soil is almost ruined already, and that small farms are being rapidly eaten up by faceless conglomerates, who know so little about farming and ecology and the soil that they are happy to blast the earth with whatever chemicals it takes to squeeze the last bit of tasteless productivity out of it before it becomes a dustbowl or a desert. Then they will move on to another commodity, like water, or perhaps CS sprays and sidearms, ready for the food riots.

* * *

There's a wise old saying from the farming world: 'Live as if you would die tomorrow; farm as if you would farm for ever.' We, the people, have the power to ensure that farming is for ever, by the way we choose to shop: the choice is essentially 'organic or not'.

A postscript to this cheerful little chapter: people like me keep going on about 'organic' this and 'organic' that, but an awful lot of people still seem to be baffled by what it means. So, if you're interested …

A Short History of 'Organic'

The problem with the word 'organic' is that it has two quite distinct, and more or less contradictory, meanings. Good old English, eh? Thus:

1. To a chemist 'organic' means 'based on carbon'. Petrol, butane, plastics, organo-phosphates … are all 'organic chemicals' because they all contain carbon. Why, you might ask, are they not called 'carbonic' instead? Good question.

They are called 'organic' because all living things, which all contain *organ*ised structures or 'organs' like leaves or kidneys, are made from this same huge class of carbon-based chemicals. So someone decided that these chemicals should be called 'organic'. An odd choice if you ask me, but you can see his point.

2. Plants and animals reared with natural fertilisers and feeds, rather than chemical ones, are called … wait for it … 'organic'. This, it seems to me, is a particularly stupid use of the word, but its derivation lies in the fact that from the ecological standpoint the animal, the plant and the soil should all be thought of *as a single 'organ'*. Well, obviously I can see the argument, but that doesn't stop it being a silly word to pick, given the obvious scope for confusion.

However, we're stuck with it, although the French use 'biologique', which would be a better choice for us as well, I think. 'Ecologic' would be even better.

The splendid Soil Association has fought to have the word 'organic' legally protected. Thus the slippery marketers and lying admen can't claim that any old chemical-raddled rubbish is 'organic'. 'Organic' now means that the crop or product has been reared in full accordance with very strict rules on animal welfare and soil conservation laid down by the SA.

A final point, in which confusion often lies: just because the chemicals that make up plant and animal bodies are carbon-based ('organic chemicals'), this does not mean that *all* organic chemicals are found in plant and animal bodies. In fact many 'organic chemicals', like DDT, dioxin, organo-phosphates, PCBs, etc., are extremely dangerous biocides (life-*killers*).

Civilisation at last

Whether or not my ME was connected with the OPs in the dip, I was still out for the count as far as the farm was concerned. To save Anne's wrists I did the rest of the rotavating. It knackered me completely and I retreated to bed for a few days.

* * *

Despite the setbacks, we were resolutely planning for the future. Little April was thriving, so our dairy supplies looked assured. We were now members of the Soil Association and were thus eligible to join the new local organic co-operative. This would mean we could sell wholesale quantities at organic prices, and be working within a supportive group, which would give us some sort of security and clout when dealing with buyers. We didn't need huge amounts of money, and hadn't come to Wales to Make Our Fortunes, or anything silly like that, but we *did* need a modicum of income to pay the phone bill.

What would our new-found security mean for us? Could we grow more crops? Or a greater variety? We still needed to improve our income by about 25 per cent if we were to get by. It could be done, we were confident of that. But how? More tunnels? Niche crops? Exciting times …

We'd also had an Improvement Grant approved, which meant that the house was at last more or less up to twentieth-century standards. That single 5amp wainscot socket in the sitting room was hacked off and replaced with five double-socket 13amp jobs. *Ten* proper sockets.* Now we could hoover in there without an extension lead from the kitchen. I

* No, not all where the 5amp job had been. They were sensibly fitted to various points on the various walls.

could listen to my music centre at last. We could spend the evenings reading in the cosy glow of over-the-shoulder angle-poise lighting and get rid of the ghastly foot-long strip lights that buzzed and strobed at us from where they'd been roughly nailed onto the two big oak ceiling beams.

Civilisation at last. Don't knock it till you've done without it.

Part of the price we paid for the improvements was a freezing spring night spent with all the doors open while the new concrete floor set. For a full month after being laid, *Cold* just oozed out of that floor. Then it settled down and became a part of the storage heater system. It warmed up, as the yard-thick walls did during the day, then slowly released the warmth overnight. This process builds up throughout the summer, and eventually the fabric holds so much heat that for the first chilly days of autumn we don't need to light the stove.

The builder? Ah, the builder. A personable young man whom I shall call Nobby. His work was mostly adequate or even good, but in retrospect I would not be happy for him to service my Lear jet. Surface gloss seemed to matter more than profound stability, if you know what I mean. Yes, 'a builder'.

My two enduring memories are of Nobby *running* along the roof ridge to straighten a tile, because it was nearly lunchtime; and of his life motto: 'You can drop me in the Amazon jungle with a socket set and a packet of three, and I'll be fine.' No doubt.

* * *

Anne's efforts paid off. She got all the crops in and watered and ready to roll. A heroic achievement.

– 5 –
Slightly Odd Rescue Day

More than one person has suggested that I make all this stuff up. I assure you I don't. It's just that a smallholding offers so much more scope for things to happen than, say, a suburban semi. Some of these things will be odd, as reality is resolutely stranger than fiction.

A friend of ours came home late one night and found his donkey curled up on the mat fast asleep in front of the fire, which was odd. On another occasion he discovered Polly, his much-loved Dorset Redpoll house-cow, wedged firmly halfway up the stairs, which was also odd. He had to get a neighbour in to help pull from the back while he himself climbed up through her legs and pushed from the front. A messy experience, as it turned out, because poor Polly panicked, with the normal metabolic result. The neighbour did not sue, which was odder.

Every smallholder has tales like this to tell. Goats loose in the pantry; dogs marooned on roofs; cats jammed everywhere from pianos to downspouts. Someone once told me about how she came home one day to find the pigs in her kitchen. Messy enough, as you might imagine, but they'd also broken into the two dustbins full of fermenting apple juice and upended them onto the floor. The wine was quite well advanced and decidedly alcoholic.

So not only was the floor a complete mess but the pigs were both completely – I'm sorry, but 'rat-arsed' is the only word in these circumstances – and had peed and pooed where they stood.

Except, of course, they were no longer standing at all. They were both crashed out on the tiles, wallowing in their own filth and the remains of the apple wine. One of them was snoring and blowing bubbles; the other one looked as though it was in a terminal coma. Should she ring for the ambulance? The vet, then? The delicatessen?

In fact, she mopped up the worst of the mess and waited for her husband to come home. Then they grabbed a back leg each and heaved the beasts out of the door and left them on the cobbles next to the hen house to be tutted and clucked over by their neighbours before repenting at leisure.

They got no sympathy when they finally stumbled to their knees, heads throbbing like tom-toms, unless you count 'You drunken *pig* ...'

We've never experienced anything quite as dramatic as any of these, but we have occasionally had oddly surreal days when The Unexpected seems to home in and scatter fragrant fragments around us.

One of these occurred a few years ago when we returned from a rare weekend away. It sticks in our memory as the Slightly Odd Rescue Day, and started with the routine rescue of a prize lamb from the pig mesh. Nothing very odd here; it happens perhaps once a year: a lamb sticks its head through the mesh and doesn't realise it can go backwards to get out. So it stays there for ever until it dies, or until rescued by a human hand; whereupon it probably dies anyway. But it doesn't happen *every* day, so ... slightly odd.

But then Anne heard a splashy scrabbling noise coming

from the other field. The only water nearby is in an old cast-iron bath, plumbed in to water the cows. It stands at bath height on quite flat ground, but close to a bank with over-hanging blackthorn hedging. Yes, you've guessed: it was a baby rabbit swimming round and round like a demented clockwork toy. We grabbed him out, pegged him up by the ears to dry (only kidding), lightly ironed him, and let him go, softies that we are. We still have no idea how he got into the bath short of effectively climbing a tree. Odd.

We then collected Dylan from the kennels and let him out for a run. He took three paces and cocked his leg. And peed. And peed and peed. And pee … eee ... eed. It just went on and on. I honestly think he'd been too shy to have a pee at the kennels, like a little boy on his first day at nursery, so he saved it up all weekend. It went on for so long that it did cross my mind that he had a haemorrhage. But he was fine. Just full of it. The third slightly odd 'rescue'.

The rest of the day was uneventful, until six o'clock when Anne saw three Yorkshire Terriers in line astern walk past the kitchen window.*

But odd things happen in suburban semis too. I've just heard on the radio of a couple who returned from holiday to find their front room trashed and vandalised to the tune of £4,000. Not burglars; not yobbos; but a squirrel who some-how got in and chewed and smashed everything in his panic to get out. You couldn't make it up.

* * *

* We've no idea who they were or where they went to. Never seen them since, either.

Of course, with a dog in residence life is never dull.

Dylan was an almost complete contrast to dear departed Porky. Where she was quiet, he was … *lively*; where she was genteel, he … wasn't. But on the other hand, while Porky was inclined to sidle away if you weren't looking, Dylan was delighted to hang out with you and be a pal. This was a great relief, as Ken next door made sure all his neighbours knew that any dogs worrying his sheep would have to be shot. Sorry, but he'd lost too many ewes and lambs over the years …

Quite right too. We never dared let Porky off a lead unless we were two paces behind her.

The dogs looked quite different. Porky was stocky of leg and had, shall we say, a fuller figure. She looked like a Jack Russell on steroids actually, finished in dazzling white with eccentric black trim, including lips and eyeliner. I've never seen another dog quite like her.

Her 'pedigree', if that's the right term, was Whippet out of Staffie.

Dylan was much more of a … what's the word? ... *cur*. There was obviously quite a lot of Doberman in him, with maybe a bit of Whippet again, and possibly just a touch of Gremlin. A visiting vet said the dozen or so pure white hairs at the end of his tail* were a sure sign of a drop of Border Collie somewhere.

But the overall effect was of 'cur': a blacky-browny beast of middling height with middling legs, a long snout, and envelope-flap ears … and a middling tail with an attractive cutlass curve to it. You'll see one or two of his relatives on

* Dylan's.

every television report from the Third World. One will be scuttling between two mud huts, with a scrawny chicken or tasty-looking sock in its mouth; another cocking his leg on a temple or rickshaw driver; another sitting in the middle of a track, gawping. We know them, every one, thanks to Dylan. Fine dogs, one and all.

Another difference between Porky and Dylan was the way they shelled peas. Porky would somehow manage to pop the pod and then lick out the peas. Dylan could do that sometimes, but largely by accident, I think. More usually he just chomped the whole pod and left a macerated mess on the floor. I guess he got an extra calorie or two that way.

But this is not to say that he was dimmer than Porky. Maybe he didn't have her natural intelligence, but he more than made up for that in his attitude. Dad said he's never known another dog that would not only look you straight in the eye for long periods, but would be clearly 'searching for meaning' when you were talking to him. Porky couldn't be bothered with all this mental development stuff, but Dylan was incredibly keen to better himself. He succeeded, too.

I've often thought that if I'd put my mind to it I could have taught him to count. It would have been a long slow job but I'm sure it was possible.

The best we actually managed between us was:

1 'We're going *out*. Fetch a stick!'
2 Ears light up, head cocks, tail rotates.
3 Rushes off in several directions: some at random (excitement and *joie de vivre*) and some specific (memory and applied reasoning).

4 Returns with stick, forcing backmost molars into it time and again most satisfyingly.
5 'Fetch another one!'
6 Gaze averted; much chomping; tail gyrating on 'idle'.
7 'Listen: no, *listen*: Fetch another one!!'
8 See (7).
9 See (8).
10 Penny drops; eyes refocus; rushes off to where he found the first stick.
11 Tries elsewhere; eventually returns with a second stick, wagging delight to the four winds.
12 'Good boy. Gooooooooood boy' etc. Much frazzling of ears.

Now then ... an interesting question: does fetching 'another one' count as a precursor to counting or not? I didn't say 'another stick', which would have been a giveaway. So did 'another' have meaning for him? Or did he just hear 'Fetch' and then bundled himself off to repeat the last successful action? After all, sticks were his most favourite part of the material world, if you don't count food.*

Yes, I guess he could just have been 'repeating' ... but somehow it didn't feel quite like that.

I find it utterly incredible that scientists ever thought that animals have no intelligence. Some still think this way, apparently. Maybe they should look in a mirror.

---

* A friend's dog would bring him a slipper when he came in from work, and would then fetch 'the other one' when encouraged. She would also bring him the door keys and drop them on the pillow if he was having a lie-in, and she was busting.

Dylan learned the usual 'sit', 'downDOWN', 'stop that', 'heel-Oh-for-Pete's-sake-*heel* you stupid ...', and all the other usual commands, including some that could be delivered non-verbally. If he was on a chair he shouldn't have been on, then a stare, possibly with raised eyebrows would be enough. A pointed finger moving towards the floor would clinch it for sure.

He understood 'Go round', if one gate was shut but another one was open. At first I reinforced this with wild waving and pointing, but after a while he clearly understood the words and would act on them. That is definitely abstract thinking, I don't care what any 'expert' says.

I saw this in action one day when we were out exercising (him, not me) in the bottom field. My labour-saving technique was to take a stick (and another one) and throw them alternately. When he'd fetched the first one back, and dropped it, I'd throw the other one. Meanwhile I'd be walking across the field so he was getting a bit of extra running in. Clever stuff, eh?

On this occasion the stick landed right up close to the pigmesh fence. Dylan hurtled off to fetch it, but when he arrived, he found it was just on the far side of the fence. He could stick his snout through and grab it, but couldn't work out how to twist the stick so he could haul it through the pigmesh. Smart, but not a genius.

He tussled at it for a while, and then I shouted, 'Go round!' He paused, then turned left and rushed off to the open gate that he knew would be the solution to his problem.

I count that as abstract thinking, and a few other things as well: having the intelligence to listen to his friend's idea;

having a map in his head of the fence and gate; and remembering that the gate was open. Smart dog.

However ... once through the gate and galloping back towards the stick, he was then distracted by a *hole* in the fence. With 'Go round!' still ringing in his ears, he ducked through the hole, then dashed towards the stick at double the speed. Obviously, he ended up back where he'd started from.

I've never seen a dog look so obviously baffled before or since. He'd 'gone round' as he knew he should, not once but *twice*: through the gate *and* the hole ... and *still* the stick was on the wrong side. My heart went out to him, as I remembered the scores of times physics had pulled similar stunts on me. The disappearing pen tops; the documents that could not *possibly* be anywhere else than where I *know* I left them; the Dictaphone that had worked its way into the lining of my coat while a notebook, coins, pens, and a conker had *not*; poor little fellow. And I couldn't explain it to him, which was even worse.

But he didn't care. I pushed his stick through for him and he trotted off, head high, tail wagging: happy in the present moment: the wisdom of the dog.

How Clever Dogs Are

Pig, I think. Slight touch of flatulence ...
Sheep ... sheep ...
Ah ... horse! unless, perhaps ...? No ... horse, and hot with sweat.
Two horses, one a male, for sure.
Sheep ... sheep ... sheep ...
Hold on a mo! Was that a whiff of ...? No ...
sheep.
Sheep ... sheep ...

A cat! A bloody cat! No … Far away and gone.
Bastard.
Another sheep. Sheep.
A ram.
… more sheep …
Hot chainsaw oil from up the Forestry
And a … a calf, I think. Brand new, at that.
What's that??? Hang on …
Aha … that's biccies, that is … dash of milk …
and topped,
unless my nose deceives me …
with chunks of fragrant and delicious
Chappie.

He'll call me in a minute. I'll rush in, wide-eyed of course,
And he'll remain convinced that I
Can understand his every word.

A Shower for April

A dog is a very good sump for pouring surplus milk into, but only to a degree. It is quite remarkable how much one pair of soppy brown eyes can say in a single glance:

'Oh please, not *more* …?'

'I've done my best for you God, but …'

'One more drop and you and the carpet will regret it.'

Etc.

So what else could we do to avoid wasting milk? The obvious thing was to buy in a suckler from our neighbours' herd of pedigree Jerseys.* 'Yes,' said Nick, 'we've got a little bull calf. You can have him for a fiver, if you like. But he'll never come to anything. You know that, don't you?'

Yes, we did know that. Thoroughbred Jersey bulls seem to be wonderfully incompetent at 'beefing up'. They just drink milk and grow bigger bones. But that was OK. At least he'd have a few months of a proper life with us instead of being abused to death in a veal crate somewhere. We called him Shower, to be a pal for April. They frolicked and bounced together very nicely. Sheer joy.

* * *

As the season progressed, we began to realise that we would soon need a new tractor as a matter of some urgency. The problem was that the TVO/Petrol hybrid (SfaS) kept over-heating. I sought advice from all and sundry. The best tips I got were to remove the thermostat, which was probably boiled to a pulp by now anyway, and to flush out the radiator with caustic soda.

I managed the first job, delighted that I didn't crack any of the alloy castings on the way, but the caustic was something else entirely … surely that poor old rad would just collapse into sludge if I abused it with a powerful alkali? I had this feeling that it was only the limescale, or whatever, that was holding it together.

And anyway, everyone knows that tractors are supposed to be diesels. For a start, they are more reliable. We needed to start looking for one, straight away.

* This was before we were brave enough to tackle pigs.

– 6 –
The Drive

Forgive me if I've mentioned this before, but one of the major hazards of smallholding in West Wales is the seasonal south-westerly. Every autumn, huge whirlpools of rain come scything in from the Atlantic with the sole aim of dumping as much water as possible on peaceable Celtic farmers. At the moment of writing we are blessed with the greatest inundation for twenty years. Four inches of rain in a day is more than the most substantial ditch can cope with and erosion and flash floods are a certainty. Fortunately, our house is unlikely to flood or erode, especially as we've had all our flat roofs replaced (*never* have a flat roof, is my considered advice), but the drive is a different matter.

Many years ago, a previous owner thought he should have a proper tarmac drive, and that the best place for it was along the bed of the existing natural drainage channel and spring line. Surely a couple of well-placed pipes would redirect the flash waters to a nearby ditch? Well no, I'm afraid not. Water, especially in bulk, knows best where it wants to go: downhill, and as fast as possible (it's in the design brief). Unless you redirect it with a great deal of forethought and several industrial storm drains, you tinker with it at your peril. Two six-inch pipes under the road just aren't up to the job when the autumn monsoons slash off the leaves. Lots of leaves block little pipes in seconds, and the roadway reverts to instant glorious riverdom.

This might not be a serious problem if the road were made of eight inches of finest concrete. But it isn't. The job was done on the cheap with a few loads of tarmac spread more or less as thinly as possible: the Marmite approach to civil engineering. The water gets between and under the stones and by morning you have potholes two inches deep where yesterday there was a surface. Once the hole has started it rapidly enlarges –and there's nothing you can do about it except wait for summer then spend £50 on 'Instant Tarmac' to patch the worst of the holes, knowing it will all wash out again next winter. And worse.

So the drive continues to flume into the yard, which is relentlessly being washed away, and we can't afford to fix it.

It does have its humorous side, however. Imagine the merry laughter and manly grit as I set out to clear the pipes for the fourth time today. 'Back soon darling. Don't wait up.'

It is dark and raining sharpened stair-rods that actually hurt when they sting against your face and hands. The drive is three inches thick in rushing water. The ditch and pipes are eight inches deeper, right up against a bank garlanded with yards of thrashing brambles. I can't hold the torch *and* use both hands on the clearing-tool (a trowel welded to a long steel tube), so the torch is switched off and stuffed into a pocket and I'm hacking and gouging away in total blackout. The rain is drumming, the ditch is roaring, and the drive is sloshing and throwing up small foaming waves around the potholes. Then my wellington, which has been suspect for some time, finally splits at the heel. The shock of a suddenly frozen foot distracts me, and my other foot goes into the ditch, which is somewhat higher than my wellington.

Laugh! I nearly did.

The moral? If a river runs through it, don't *ever* put a road there.

* * *

Water is a constant source of minor irritation to us. I know it shouldn't be, and that we should be grateful for it when half the world is desperately thirsty, but you can have too much of a good thing.

At the very moment that my second wellie was filling up to mid-calf I recalled that out here in the country people pay dowsers good money to find water on their land. 'Oh ha ****** ha,' I thought, as my toes turned to tiny little super-saturated icicles.

It's true though.

If you need to water cows in a distant field, then you can either run a PVC pipe out there, which would be a long and expensive job, or you can find a spring and drill down into it to fill a little pond, or more commonly an old bath tub.

So how do you find the spring? Well, you could employ a hi-tec ground radar specialist to drive up and down all over every yard of the field, and who may find nothing anyway, or you could have a word with Gwyn, who will turn up with his antique hazel twig or whatever he fancies and follow it round until he's found the optimum spot. He can also tell you how deep you'll need to drill, and the rate of water flow.

Dowsing is as old as mankind, and nobody knows for sure why it works. Science, unfortunately, tends to view the phenomenon as merely quaint and unworthy of serious attention. Yet to watch a dowser's twig at work is to echo Galileo's famous words when the Church told him the Earth did *not* go round the sun, despite all the evidence that it clearly did: '*Eppur si muove.*'*

Dowsing is not limited to finding water, either. I once stopped and chatted to a man with two dowsing rods walking slowly over some waste ground in the Black Country. He was working for the Electricity Board and was looking for old underground cables.

People who have tried it say you can dowse for just about anything. It might help if you hold a sample in your hand (electricity?) but it's not vital. What matters is that you are very clear what you are looking for and allow the twig or rods to amplify your own slight muscular responses.

I'd read somewhere that there are such things as 'black lines', associated with some kind of water-related energy field, which can have a debilitating effect if you are exposed to them for too long; and our house does lie on a spring line. Maybe 'black lines' might have something to do with my condition? Any cure or alleviation, laughable or otherwise, would be better than none.

A friend suggested we ring Daphne, who would be pleased to check out the house for us ...

What were we to expect? A black-catted crone? A diaphanous nymph, trailing drifts of stardust? A vicar's wife in a shift and plimsolls and a loud braying laugh?

Daphne turned up with the tools of her trade: a pair of right-angled wire rods, which she held out in front of her, and a wicker basket containing a pretty gemstone (a peridot), a

* Translated as either 'And yet it moves' or 'Oh yes it does, *as you well know.*'

sheaf of rusty nails in a pouch, and a lump hammer. She walked slowly past the house. Suddenly the rods crossed. She marked the spot. One metre further on, they crossed again. Then again. Then again.

To cut a long story short, Daphne reckons most 'black house' problems are caused by four or five 'black lines' at most. We had thirty-two of them; nineteen swept in from the north, and thirteen from the west. They met in a grid pattern underneath the house. We were living in a crossword of negativity.

The answer to a black line is to 'tie it off'. Daphne checked she'd found the exact location of a line using the rods, then placed her peridot on it. To the north of the crystal, the rods crossed as before. To the south, they did not. Bingo.

The next step was to bang in an iron peg in place of the crystal, and lo! the black line was tied off.

Barmy? Well … Daphne and her assistant both got consistent results, which was interesting. Then Anne discovered she could do it too, and her results were also consistent.*

Barmy or not (and the idea of television or flying to the moon was once thought 'barmy', as no doubt was using tools or standing up on your back legs), the point is: 'Does it work?'

Well, oddly enough, there is a sign that it might be working. I wash up in the mornings, and have often felt drained half-way through. Anne dowsed the kitchen and found a crossover point right in front of the sink. Since she tied it off I've no longer felt bad while washing up. A coincidence?

* I could *sort* of do it, but was not very consistent, which led to Daphne calling me 'weird'. 'That's one for the diary,' I thought. 'I've just been called "weird" by a woman who's walking up and down in front of my house with a pair of minimalist six-guns and a bum-bag full of rusty nails.'

All in the mind? Let's wait and see. ME notoriously recedes and returns.

* * *

I wish I could report that Daphne cured me, but alas, the evidence is 'inconclusive'. Up and down, as ever ...

However, I've tried a bit more dowsing myself, and even went on a course. We used the L-shaped wires made from old coat hangers, and held pistol-style. 'Go and find the drain in that field,' the teacher said. And lo – the wires did cross. No doubt about it. But only sometimes. And not necessarily where other people's did. Interesting though.

I recently tried a bit of emergency dowsing to find a wad of papers I'd mislaid. I made a pendulum from a half-inch steel nut hanging on a thread of split baler twine, drew a (very) rough map of the house on the back of the traditional envelope, and held the nut over it. 'I'm looking for my Talk Notes. OK? Not the drains. Not a buried cable. Not gold. Not magic beans. *My Talk Notes.* OK?' I dangled it over the plan of the upstairs rooms. 'Are the notes in this room?'

One has to negotiate with one's pendulum beforehand, and agree on a code. For example, a clockwise rotation will mean 'Yes', a counter-clockwise one will mean 'No', and just hanging there, like a bell-less clapper, means ... well, I'm not sure exactly, but Dumb Insolence will come into it somewhere. Or maybe you're asking the wrong question. Anyway, the solution is to keep asking various questions until you get a response.

No, my notes were not upstairs, which was comforting, as there was no conceivable way they could be.

I then hovered over the less likely downstairs rooms: Utility? No. Kitchen? No. Bathroom, for heaven's sake? No.

My confidence was building. Before long there were only two rooms left. 'Back Office?' (... where the notes really ought to be as I'm *sure* I put them back in the drawer ...) No.

Oh.

So that only left the Computer Suite (my parents' and then Paddy's old bedroom which I now use for writing in and storing large stacks of rubbish). The notes *must* be in there ... My heart sank ... piles of stuff everywhere. It would be impossible to find the notes, even if I'd tried filing them sensibly, because I can never understand my own filing system. Once something gets filed it's gone for ever.

'Are the notes in this room?' and I carefully hovered the nut over the drawing. No response. 'Oh for Pete's sake ... we've eliminated everywhere else. At least be consistent ...' and slowly the nut began to describe a small circle ... *anti*-clockwise: 'No.'

Oh spiffing.

I briefly considered hurling the nut through the window but thought better of it as it would probably only get lost in the rockery and I'd never find it again, unless I dowsed for it.

Then I realised that I'd omitted to draw the front hall on my map. This is a tiny cul-de-sac leading to the ex-front door. Defunct for decades, it contains only books, a toy panda on wheels and a full-size model of Felix Fénéon, the French art critic.*

'OK, Oracle ... are my notes in the Hall?' A tiny clock-

* No, really. More later.

wise movement by the pendulum. Heavens, this was the first clockwise response.

'As I face the old front door are they on the left?'

Yes.

'Honest?'

Yes.

A small gulp. Time to go and see.

I've fitted ten bookshelves in the hall, and there are books and Stuff Assorted piled everywhere. Let's start with 'the bookshelves on the left' ...

And straightaway ... there they are. In their plastic wallet, on top of the books on the shelf at waist height, but concealed from view by *A Brief History of Time* and the *Pop-up Kama Sutra.* I'd obviously come home tired and carelessly 'shelved' them in a place I would never have deliberately chosen, to be sorted 'tomorrow'.

So what do you make of that?

The new tractor

Now then ... that new tractor. First of all, let's be clear that 'new' in the circles we were mixing in meant 'second-, third- or possibly eighth-hand'. 'Does it work, more or less?' is the key question. If 'No', then 'Can it be fixed, *really* cheaply?' (We'd already done the sensible thing that hard-up farmers do and bought in a complete Fergie wreck which could be plundered for spares. It was currently reposing in a huge clump of brambles, waiting and rusting.)

'... And how on earth are we going to afford it, whatever?'

We were driven by necessity. We did need a tractor, even

though we didn't really want one (SfaS) and we needed it to be reliable and not need hours of cuddling every springtime, as our TVO model did (SfaS). It needed to be a *reliable* diesel.

I remembered Alun the Mechanic's golden advice that 'nobody ever sells a good van', and vaguely realised that this would undoubtedly apply to diesel tractors too. But the ME was still sloshing round my brain and straight-thinking was beyond me. I was a mug in the making.

We scoured the local papers for weeks with no luck, but then one filthy February morning we found an old Fergie Diesel advertised and drove out to see it.

The vendors couldn't believe their luck. Did I have

MUG PUNTER OF SARON, PARIS, NEW YORK

stencilled on the side of the Reliant? I might as well have (had). They watched warily as we got out of the car. One of them adjusted his greasy flat cap and said, 'The tractor, is it?' I nodded. His son pulled at his own flat cap and grunted, so we followed him.

Round the corner was the filthiest tractor in Christendom and far beyond. Oily this, oily that, bent everything. It actually seemed to be a shade knock-kneed. It was the most broken-winded and spavined old beast of a tractor you could imagine. But it was a diesel. And it was available. And we definitely needed one.

That was all I could think of. They wanted a silly price for it and I don't even remember haggling. This was not only bad economics; it seriously breached a bedrock law of the countryside. You haggle over everything. It's absolutely expected. Then you agree on a price in the middle, as expected, shake on it, or slap palms, then return to the purchaser a penny or a pound as appropriate, 'for luck'. Then you're on your own. No such thing as a guarantee, obviously.

They were so shocked at getting far more than they were expecting that they threw in an old cultivator. This is another splendid piece of Fergie kit that works off the hydraulics: two bars with a dozen or so curved and spatulate spikes spaced along them. You adjust the position of each spike, then drive between your crop rows and the spikes do your weeding for you. I guess you could also use it as a sort of rough and ready 'plough-tavator'.

Deal done. Home. Anne took the car and I took the moribund hack. It had started easily, which was virtually the only thing I needed to know about it. The gear stick waggled round like a spoon in a sundae, but all the cogs seemed to be there, and the brakes even worked, which was an unexpected bonus. It was definitely good enough for our purposes. We didn't need anything big, posh, supercharged or sophisticated. Just something that would do a bit of modest ploughing, harrowing and occasional carting, and then *start* next time it was required.

As it happened, there were three proper Welsh hills on the way home: two up, one down. All steep enough to need a driver's full attention.

We pulled slowly up the first one. I'd been up this hill in the TVO tractor and it had roared up it. Plenty of power. But this poor old diesel dobbin didn't. It wheezed and rumbled and *groaned*, all the way to the crest. Oh dear.

Going downhill was fine. It could obviously do that all day, especially as the brakes were properly balanced, with only a token kamikaze twitch into the oncoming traffic. But I was becoming aware of a definite smell of sizzling diesel, as the engine braced itself against the gearbox.

The second uphill was the steep one. I'd once run out of fuel halfway up this brute in the Reliant, and had posed a considerable nuisance to all and sundry while I fetched Alun. In fact I'd not run out of fuel. It was just that the filter was blocked. He whistled through it and put it back. Vroom …

We did reach the top, at a slow walking pace, where Anne was anxiously waiting. 'It sounds like an old tank,' she said, smearing my glasses for me with a woolly scarf.

Ah yes. I should point out that it had begun to drizzle within minutes of hitting the road. Just drizzle … but persistent and cold, and the heater didn't work.* And it was February. My hands were rigid on the steering wheel and my nose was running, but the plastic trousers, wellies and my home-waterproofed coat were keeping me mainly dry. However, I was so liberally basted with diesel that I definitely constituted some sort of fire hazard.

A terrible mistake, that tractor was. It was oozing oil and lymph from every pore by the time I hauled myself off it, and stood immobile in the yard, hunched over, hands set in rigid claws.

'Cup of tea?'

'Hrnghphnm …'

'Is that a "yes"?'

'Eph fhlee … z'

* No, of course there isn't a heater on an old tractor. I'm making that bit up.

– 7 –

Atoms Deep, Genes Deeper

The negative responses to genetically modified foods seem to me to fall into four categories. Firstly, people are concerned that it might poison them. I don't know much about how DNA is metabolised, but my impression is that direct poisoning is not really an issue. However, there was a recent report that GM pollen had killed large numbers of Monarch butterflies in the USA.

The second concern is that GM plants pose an unknown and possibly catastrophic threat to world ecology. The chemical companies claim to have done their homework, but do they know what effect *any* grain of GM rape-seed pollen will have on *any* charlock weed? What properties will any resultant hybrids have? And how will they affect the insects that pollinate them (remember the Monarchs)? Or the birds that eat those insects? Or the fox that eats the birds? Does Global Mutations Inc know all these answers? No, of course not.*

And what sort of plant might emerge when GM pollen meets a natural mutation in a charlock, or indeed, the rape

---

* Just eighteen months after writing this article, this precise crossing of rape with charlock has been reported, although thought to be 'virtually impossible' by government scientists. There is thus now a herbicide-resistant 'superweed' spreading across Dorset. How many more of these 'virtually impossible' yet utterly predictable crosses can we expect?

itself? And what about backcrossing, from polluted charlock to rape? There is no way these questions (and many others) can be answered in advance of the event itself. The results may be quite harmless. Or one in a million might be voraciously poisonous or overdominant in a finely balanced ecosystem. And let's not forget there is only *one* ecosystem, and that pollen can travel enormous distances. Every scientist knows this, so I find it baffling that so many behave as though little bits of The One can be somehow isolated and fiddled with, with impunity.

Thirdly, people are worried that over-reliance on GM seeds puts too much power in the hands of Big Business. We remember how the Green Revolution, which promised a superabundance of cheap rice in Asia, collapsed when it was realised that the super-rice varieties (supplied by Big Business) required superdoses of very expensive artificial fertilisers and pesticides, also supplied by – guess who? Yes, Big Business.

But it seems to me that the passion of the public response to GM indicates a reaction at a yet deeper level, which is (as yet rather inarticulately) asking 'Do we know what we're tinkering with here? Are we getting out of our depth?' And the 'we' means 'scientists', because it is they who are doing the tinkering.

Our attitude to the role of science in society is slowly changing. For a hundred years, it could do no wrong. It brought us electricity, medicines, mass transport and communications, and an escape from drudgery for millions.

But in the 1960s attitudes began to change. People dared to suggest that just because science said something *could* be done this was no longer a good enough reason why it *should* be done. Concorde was fast and beautiful, yes: but highly polluting when balanced against its effectiveness. People protested. Nuclear power was also questioned. Yes, it was clever, and smokeless: but what about the huge hidden costs and pollution?

A series of other 'minor' scandals followed in the 1970s and '80s: some down to government; most down to Big Business; *all* developed by science, which is what the public remembers most. Remember thalidomide, and the mutated births? Acid rain? The deaths from dioxin at Bhopal and Seveso? The poisoning of the Rhine by endosulfan? DDT that made mothers' milk unsuitable for consumption? PCBs? OPs? Even in the 1990s a full half of all wood pigeons still had DDT in their fatty tissues, some ten years after DDT was banned.

Public reactions have included establishing CND, Friends of the Earth, Greenpeace, and dozens of other groups pleading for responsible uses of science; and me, and possibly you, going organic.*

Each new alarm makes people more suspicious of The Experts, who, in a techno society, mean The Scientists.

It is this growing trend of disillusionment with science-as-panacea which is behind the GM panic, I think. People now know that nuclear power, which was originally promised to be 'too cheap to meter' has indeed proved to be extremely expensive, extremely dangerous (Chernobyl; rotting Russian submarines; the endlessly leaking Windscale/Sellafield/

* Just as a matter of interest, organic farming actually helps to lock carbon up in the soil: it is thus an active force against global warming. Three guesses whether chemical farming does this ...

Cuckooland reprocessing dump, etc.), and polluting beyond belief.

Rutherford famously admitted he did not know whether he would blow up all Cambridge when he split the atom; and he was a *responsible academic*. How far, people wonder, can we trust the reassurances of Big GM Business whose motives are not wisdom but profit? And, even, how far can we trust our 'independent scientific experts', who tell us there's no problem? These same 'experts' told us that BSE wasn't a problem, and that GM rape wouldn't cross with weeds. People remember these things.

I guess the big issue here is whether or not science will continue to dominate Western (and increasingly global) thought. Or will more individuals and even governments begin to question its overwhelming influence?*

* * *

DNA has been hailed as 'the secret of life', which scientists are quite competent to manipulate as they choose. But DNA is just a chemical: un-alive, by definition. What it does, as I understand it, is to act as a template in the hugely complex system of protein assembly: a process that only occurs within a cell. No cell, no protein assembly.

And the cell must already be alive. Dead cells do not assemble proteins. DNA is a chemical tool used by a living cell. *So, to understand DNA, we must first understand Life, and*

---

* Since the nineteenth century, Scientific Materialism has replaced Church Idealism as the ruler of the socio-political roost: one ill-considered and aggressive belief system has replaced another. Plus ça change.. (SfaS)

not the other way round. It cannot be 'the secret of life' or anything like it.

The properties of DNA are actually turning out to be a huge mystery. Simplistic hopes of finding a gene for crime or a gene for genius are being quietly abandoned. In fact it has been discovered that man, the thinking, conscious, would-be manipulator of planets, shares about a third of his genes with the common or garden daffodil, which just goes yellow, then stops, like a lightbulb. What we should make of this, I have no idea, but I would imagine it might be a bit embarrassing to the Materialist-Mechanical model of Life. Maybe it's time for a rethink?

* * *

To bring the GM issue back to earth … what of the poor organic farmer? He is trying to grow honest-to-goodness food of the sort that doesn't give children asthma or turn teenagers into destructive little jerks, with very little real support from the government, or anyone else.

For decades he's had the problem of what to do about any pesticides that his neighbour might allow to drift across his land (not a trivial problem, by the way). Now he's got to worry about the impact somebody else's GM pollen might have on his own crop. It might come from scores of miles away.

All in all, once released, GM is for ever and Pandora Plants™ will rule our world.

What will we choose? (The choice is yours, of course, dear reader: what will you choose to *buy*?)

Woodcraft Folk

Spring sprang, and the land came back to life. Seeds shot up, as amazingly as ever, and began to live out their purpose. Anne nurtured them and they prospered.

I began to rouse, as well. The summer was approaching and the crops needed all the help they could get, particularly with weeding. 'Weed it and reap', as the saying has it.

It seemed such a shame to waste all that balmy weather and those beautiful views, and as we were always on the look-out for some small and legal way of boosting our income, we thought we'd try a new crop: campers.

We fenced off the top section of the bottom field, and reckoned we could fit in half a dozen caravans and a few tents without overcrowding. However, we might have to build more 'facilities', unless all our visitors fancied a hundred-yard stroll, then a scramble over a barbed wire fence before hacking out a suitable niche in a clump of nettles on the edge of the cwm; and then possibly hurtling to their doom in the stream below, with well-prickled privates.

The snag? Advertising. It costs a packet to advertise in magazines, and you have no way of knowing whether it's going to pay off. We just didn't have enough money to risk on regular advertising, but we did try it just the once.

As a result one quiet couple in their twenties turned up in a Land Rover and asked if there was room for them. I didn't laugh. They parked bang in the middle of the patch and were quiet. I told them where the loo was and they said thankyou. Would they want any fresh veg? No, thankyou, quietly. Milk? No thankyou. Er ... well, should I do my local dance, jetéeing from cowpat to cowpat? No thankyou.

Next morning, they paid their modest rental and left, quietly.

Income (rent) over expenditure (advertising) was about minus fifteen pounds.

The other happy camper was an old friend of Dad's who brought his caravan all the way from Kent. I have a photograph of his car, with its Caravan Club sticker, right next to the house and the Dwlalu Farm nameplate.

Did this impress the Caravan Club when we applied for an official entry? No. The access was too narrow, they said. Had anyone actually been down the drive to check? No.

We did have one much bigger group of campers: a weird tragic-comic experience that involved a car crash, the Co-op, a mercy dash by the miners of the East Midlands, and a group of Portuguese teenagers. And a missing cat. And rain.

No doubt you will already have worked out the story for yourself, but just in case you haven't, this is roughly what happened:

Back in Nottingham, Paddy had belonged to the Woodcraft Folk, a sort of left-wing alternative to the Boy Scouts.* He'd enjoyed it and we'd stayed in touch. One evening we had a phone call. How would we like to be the venue for the Nottingham WF annual summer camp? There would be a dozen or two kids, plus several adults, oh, and a visiting troupe of Portuguese WF as well.

Excellent! It would be good to see old friends again, and

* I am not interested in debating who is or is not left or right wing, by whatever definition. Go and pick a fight with somebody else.

all those people would need lots of milk and veg, and possibly a fatted calf or two.

The advance party turned up in a borrowed Co-op removal van, late one soggy Friday night. They didn't bother trying to negotiate the narrow access by the barn, and came straight down the top field instead. A sharp left turn (in a removal van?) and they hoped to be in the camping patch. But the field was sloppy, sloshy, squishy and sodden, and removal vans are heavy. They didn't even reach the bottom of the slope ...

The plan was that the womenfolk and the kids would arrive the following day at Carmarthen station, where a coach would collect them. Smiles and merry song were guaranteed.

At 7am we were woken by a call from the police. There had been an accident and several people were in hospital. Oh lor' ...

Joe had set out with four kids and his trailer-tent in the early hours and had come to grief on a bend. We roused Glen and Dave from their lumpy slumbers in the removal van.

The phone earned its keep that morning. Calls flashed back and forth to and from folks in Nottingham, the hospitals, and the police, but hard information was hard to come by. Glen rented a car so that Dave and he could visit everyone concerned.

This left the camp base unmanned all day and the women and kids were due late afternoon. Of course there were no tents up, and of course it was now raining again.

At tea-time the coach dumped all 30+ of them at the top of our long drive and they shuffled in like a line of refugees, travel-worn and damp, hauling their bags and cases.

Fortunately our Big Black Barn wasn't yet full of hay, so

it served as a base for emergency cups of tea while The Plan was sorted out. There wasn't a lot of discussion as all The Plan could possibly involve was: get that flipping kitchen tent up a.s.a.p. All else would follow.

So a chain of people ferried the bones and pelt of the two-poled kitchen tent out of the van and got it up. The ground was super-saturated: puddles everywhere, with rivulets snaking through the sward and rushing down to the stream in the cwm, which was already roaring like a train.

Then it stopped raining and Glen returned from his travels. The good news was that nobody had been killed.

'But where's Dave, Glen? Did you leave him at the hospital?'

'Er, no. He's with the car.'

'And where's that?'

'In the ditch at the top of the drive.'

Anyone with half a muscle went to get the car back on the road with the aid of lots of wooden chocks and slats and some impressive shouting and arm waving. Never fails.*

By the time it was dark everyone was more or less sorted and somehow the womenfolk cobbled together baked beans and sausage for everyone.

The cask of essential beer took a bit of a bashing and then Jean and Glen took their kids (including a babe-in-arms) to kip in our caravan while everyone else spread themselves out in

* I'm amazed by how many people go into that ditch. The road is perfectly wide enough, even for a removal van (obviously) but every six months some hapless car or van plunges into it. I can only imagine there must be the local equivalent of Sirens hovering over the tarmac: 'Plunge! Plunge off the perfectly adequate road and into the ditch. Do it now...whoooo...'

either the barn or the kitchen tent and prepared for refreshing sleep and a new day.

The rain came back, good and heavy, and to add to the fun the RAF saw fit to hold a vigorous night training exercise directly above our barn and occasionally straight through it. The noise was tremendous, and mightily amplified by the corrugated iron. Every time a jet screamed over everyone ducked and shuddered, except for Dave. 'Personally, I don't mind them,' he explained. 'It's the RAF that enables us to sleep quiet in our beds at night.'

We had organised a back-up plan: the local minister had offered use of the chapel hall for the night, if needed. Thanks, brother, but a barn is fine, even if it doesn't have a piano.

At 2am on Saturday morning the cavalry arrived. Anne and I were woken by the roar of engines and the flash of headlights on our bedroom ceiling. Eh? Injuns? Martians?

Actually, they were Woodcraft Folk leaders and supporters who'd driven two hundred miles from Nottingham after work to help out colleagues in distress. They kipped in their cars, naturally.

Next day they helped sort out the tents and most importantly rescued the van, still up to its axles in our field.

How did they do it? Alan, a Calverton miner, had been a desert soldier and knew a thing or two about rescuing bogged-down trucks. Using only a couple of marquee guy ropes, two big marquee corner pegs, and a pair of spanners, he winched the removal van slowly backwards, out of the mirey slots the back wheels had cut, and back onto firm(ish) land. Brilliant. No tractors, no helicopters.*

* Have you worked out how he did it yet?

God knows what the Portuguese made of it all. 'One for the memoirs', perhaps. Or maybe 'Hello Muddah, hello Fadduh, wish I was in Camp Granada …'

The rest of the Camp went smoothly, if somewhat damply, and people even began to enjoy themselves a little.

Eventually the week passed, and everyone said goodbye and packed up and headed back to Nottingham, and that was that …

… until that evening, when we got a phone call from Glen to ask if we were missing anyone. No? How about Pudding? They'd been unpacking the marquee and found our dopey cat gawping back at them from inside a swath of canvas.

We got her back the following week, as another visitor kindly collected her from the Woodcraft Folk and brought her with him.

That was our last venture into camping. It clearly was not meant to be for us.

* * *

The crops thrived and although we'd lost all the garlic the previous year, we somehow managed to sell enough produce to stay alive for another year.

All in all, we were very pleased with 1986. We'd survived a crop disaster, and joined the co-op, which would give us some much-needed stability and help with the endless headache of marketing. And we'd booted the ME into touch.

Or so we thought. But in the middle of December my nose began to drip again, and lassitude returned. Back to bed.

– 8 –

For 20 Minutes, Read 190

'It'll only take twenty minutes. No sweat.'

I should have known better after thirty years of DIY and well-intentioned efforts at improving my cash-saving craft skills.

My projects are not particularly ambitious. I have never set out to move the farm slightly to the left, or erect a motte and bailey on the lawn. I aim only at modest repairs and improvements.

And it's not that I am particularly incompetent. I very rarely make a complete pig's ear of anything (except silk purses, obviously); although I did once carefully measure a length of carpet, and then carefully cut it perfectly square and true, precisely three feet too short. Oh, we did larf.

And I'm actually rather good at decorating, though I say it myself.

So what's my problem? Hard to define, precisely – it's just that despite all my efforts at thinking the job through and getting materials together in advance, nevertheless, bizarre and foolish hitches delay a perfectly simple job, time and time again. It's as if the God of Small Repairs sees me getting my wrenches and screws out and settles down for an afternoon's fun with a bowl of swine scratchings and a six-pack of ambrosia. 'Yo, Zeus! Bit of a giggle coming up in West Wales. Pass it on.'

This morning, for example: I have a small cantilever

desk-thing whose varnished top has been so abused that it looks like it's been smeared with a thin ripple of oven-baked vindaloo. What could be easier than to (a) sand off the old varnish; and (b) apply a new coat? Twenty minutes, max. I assemble a variety of sanding devices, the varnish and a brush, and set happily to work in the August sunshine. What could possibly go wrong?

For the first fifteen minutes, nothing. The belt sander strips off the crusty varnish like creamed rice and apricot off a baby's cheek. I step back to admire the finish. Suddenly, rain from a clear sky. Splat. Splot. Just four drops ... onto the bare wood. OK; a trip to the bedroom for the hairdryer and a ten-minute delay. No sweat.

Nearly ready to paint. Just one more sweep with the sander. And it oozes a blob of black grease onto the plywood.

Into the kitchen (the toolshed; the bedroom; the tractor barn) for the white spirit. Find it in the Nasty Place Under The Stairs where I keep my tins of dried-up paint. Another twenty-minute penalty.

So: the sanding is done, all bar the bevelled edge. There is no future in using the belt sander on this as it would only end in tears and a flayed arm, so I apply my little Sandvik device: basically, a handle attached to a metal plate with an orange-peel finish. Excellent for sanding bevels.

Three inches into the job, this intensely complex machine goes wrong in the only way it can. The tin plate drops off the handle.

At this point Victor Meldrew takes over, and memories of previous persecutions surface. The doorknob that fell into two pieces in my hand; the nut on my daughter's bike that *not one* of my *seventy-four* socket, ring, or open-ended spanners would fit; the log that trapped my saw *and* two chisels in its twisted veins.

But ... I have Evostik, and I have clamps! Thirty-minute delay. Try again. The tin bit drops off immediately, and sticks to the floor.

Glass of homebrew; deep breath, remembering that the purpose of the Universe is not simply to embarrass and humiliate me; it is there to embarrass and humiliate everyone. So why do I take it so personally? And why am I so convinced that if I placed a six-inch cube of lead on the blue spot, it would immediately roll into the corner pocket? More deep breaths; long but vital delay talking it through with the dog.

Then a quick lick round the bevels with a sandpaper block, and we're ready to paint. No sweat.

All I need is a screwdriver to open the lid with. To the toolshed. Open the door. And the side of the doorframe with the sneck on it falls away from the wall, onto my array of fixings, and sprays Screws Asstd and Blue Tacks all over the floor.

I am beyond anger or persecution by now, and just walk straight past, Stan-Laurel-style.

The finished product looks fine, and is much admired ('That's nice dear'). It is marred only by three lacunae caused by brushing through a drop of perspiration which dripped because my bifocals make a mockery of normal vision for close work unless I lean right over it.

So. Job done. It took a little over three hours. No sweat. Well, almost.

* * *

Not every job takes five times longer than it 'ought' to. Some just sail through: making a chicken bivvy from a couple of sheets of corrugated iron and some bits of left-over battening, for example. I was astonished at how quickly it just knocked together. Perhaps the God of Small Repairs was having a day at the seaside, sinking little children's boats and pushing old ladies down concrete steps.

On the whole I am very safe with tools. I have only two scars to show for half a lifetime of bodging. One is an inch long and on the ball of my right palm. This was caused by a slipping screwdriver. I was holding a suppurating section of Reliant intestine in my right hand while trying to loosen the screw of the seized-up jubilee clip with my left (yes, I *am* left-handed, and not merely stupid). The driver skidded while I was pushing hard on it and skated off the clip and straight across my Mount of Venus,* opening a deep and bleeding channel en

---

* Whoa there! It's a term used in palmistry.

route. The moral I drew from this injury was: whereas the best way of avoiding a cut is to use a very sharp knife, the same does not apply to screwdrivers. Make sure your screwdrivers will *not* slice a sheet of tissue paper before starting work.

The other injury did involve a knife, and a very sharp one at that. While the ME was upon me I was not capable of proper manly work like toting barges, lifting bales, and so forth, and was relegated simply to doing the most useful job possible. Some days this involved gawping at a wall and other days I might shunt a hoe round a polytunnel for half an hour while Anne got on with the real work.

On this particular day the top job was courgette trimming. Anne was out on the field, doing the actual harvesting from 3,000+ plants, then bringing an endless line of over-stacked three-gallon buckets into the packing shed for me to deal with. I took each courgette carefully out of the bucket (mustn't bruise it), checked it over for damage, and then classified each fruit as either 'Grade 1' or 'Other'. 'Grade 1' could mean either 'six inches long' if they were destined for Tesco, or the much-preferred 'seven inches long' if going to Sainsbury. 'Other' meant anything that wasn't 'perfect' according to the supermarkets' lunatic, random, and mean-spirited ideas of quality. 'Others' went into a box which the local fruit and veg man would sell for us at a much better price than we ever got for our supermarket 'Grade 1's. Make sense of that if you can. (We didn't send *all* the courgettes to the local man simply because we had too many.)

Some fruits needed their 'picking-ends' to be lightly trimmed and squared off, because otherwise a supermarket was liable to reject the whole batch. Just one fruit *slightly* less than 'perfect' according to some ignorant Suit's idea of quality … and they'd reject the whole batch.* The courgettes wouldn't keep, of course: so a dead loss. All that work for nothing. So we needed to make sure every end was neat and tidy. Hence the very sharp knife: one of those snap-off jobs, sharp as a councillor's wit.

I'd been trimming and packing for a couple of hours, when my concentration began to fade and wander, and I was looking forward to sitting down and staring at a wall again. Just a few more …

Suddenly, 'Hello? *Red?* What's that red stuff? Oh …'

I'm no expert on blood, but I know it when I see it. And I knew it could only be mine. 'Right. Where's it coming from?'

Common sense says, 'Well, it's not going to be your liver or your toenails, is it?'

'No, probably not. Point taken. Ah …' There it is … The tip of my index finger appears to have become unaccountably flattened and to have turned into a scarlet oozing pad, suitable for signing Japanese prints. 'Finger flat ... finger bleeding ... more than a scratch. Call help.'

Anne stumbles in, attracted by my calmly measured shriek. Bless her, she's brought two more buckets with her.

'What's up?'

I show her the flat red tip of my finger.

'Where's the top?'

* Even if the damage was nothing to do with you. For example, a supermarket might employ a contractor to (pointlessly) wash your spuds or carrots. His equipment might cause some minimal scarring, leading to rejection of the batch. But it's the farmer who pays for the ensuing dead loss. Bet you didn't know that.

Ah ...

We poke about among the courgette trimmings and find the bit of finger. It fits. Yep ... definitely mine. It won't stay in place though, and we both realise that this job will need more than a bit of Elastoplast.

Anne rings the hospital and tells them her husband has chopped the end of his finger off and can they help, please?

'Bring him straight down.'

Anne drove the thirteen miles to the hospital while I kept my finger upright so the makeshift Sellotape scaffolding would hold the tip on more or less square. We didn't use sticky plaster as that would have been too permanent and thus involved too much eye-smarting and scream-inducing rrripping later on; then all the messy business of trying to prise the tip off it afterwards.

Straight into Casualty, dripping onto the nice clean lino. My hanky was a sodden pirate flag.

The lady on the desk smiled: 'Ah ... are you the gentleman who's shot the end off his finger?'

'Shot?'

'You said on the phone.'

'Chopped. She said "chopped".'

'Oh … well that's a relief. We were wondering how you could shoot the end off your finger.'

Good point. A shotgun would take off most of your arm. What else? A rifle? How could you reach? And why? A pistol? Ah …

What would I be doing with a pistol? And did I still have it in my pocket …? I was looking a little wild around the eyes. And I *did* have a beard …

'Chopped' was a relief all round. A tidier wound, too.

They put me into a cubicle where we waited for a doctor. By now I was feeling a bit squeamish as well as wild-eyed. Delayed shock, I guess.

The doctor was Arabic, I think. He held my finger up carefully and looked all round it. Then he prised off the Sellotape; then the tip of the finger. 'We won't be needing this, I don't think.'

'Hey, hang on,' I thought. 'Who's this "we"?' The best I could manage was a wan smile.

'No. The tip will grow back quite nicely on its own, almost as good as new. If we try fixing this back on it will only invite blood poisoning.'

Smiles all round. A nurse gave the digit a quick wipe over with an oily rag and a dash of talc and that was that. A big puffy bandage. Done. No lollipop, however.

Fortunately, the injury made very little difference to my work rate as I was doing very little work to start with. And it's surprising how little a damaged index finger matters, unless you're an aggressive politician or a disco dancer.

My only regret over the whole affair was that I didn't think to ask if I'd be able to play the piano afterwards ...*

The Year of the Cow

The winter passed in a haze again. I was a complete waste of space, which was a worry because the smallholding demands two adult full-time workers *all* the time. Our doctor was very

* I won't spell the whole joke out. Ask someone else.

sympathetic, but could offer no real help. Seasonal Affective Disorder (SAD) was suggested and rejected. I forget why.

We joined a couple of ME self-help organisations and Anne pored over their publications looking for anything helpful in the way of cures, alleviations, or even causes. Nothing. Every study showed that every 'cure' turned out to be no such thing after a period of proper testing. Or, worse, it worked for some people but not others ... for no apparent reason.

My own contribution to this process was to completely ignore it. For a start, when your head is full of blancmange you can't understand long words like 'tiresome' and 'captive bolt'. I also hold to the (predominantly masculine) school of thought which says that to study something is to condone it. One way and another, I didn't want to know. It was clear that medical science hadn't got a clue, so what was the point of wasting my precious time trying to read about it? I was surely better off doing something better with my odd periods of semi-lucidity. I began to re-discover reading: science fiction at first, then general literature.

As had happened the previous year, I began a slow recovery in the springtime.

* * *

1987 was a Year of the Cow for us. If you are truly intrigued by bovine genealogy then by all means pay close attention to the dates. Otherwise skip swiftly over them. They don't matter.

Daisy had been our first, bought the year after we arrived on the farm. She came with a bulging udder, as her calf had just been weaned off her. Gallons and gallons of the richest milk in the world. We made a lot of butter (SfaS).

Then, as is the custom, we put the AI man onto her.* This 'servicing' is normally done in July, three months after the traditional spring birthing. As we wanted a replacement for Daisy, we specified a Jersey bull as her suitor.

In the casino you get either red or black. In breeding you get M or F. We got M, and called him Wally.

There was a problem, however. Daisy looked a bit off colour the day after the birth: sort of dispirited, and not interested in her hay, or even her double-scoop of 'sweeties' (concentrates: molasses and beet pulp with added vitamins and, since the BSE scandal, probably no diseased sheep brain).

She suddenly began to look listless, and well ... *pale.* We felt her ears ... they were cool. While we were debating whether to bother Ken again so soon after he'd rescued Wally from an early grave (SfaS), she began to sway. Seconds later she staggered, stumbled, and then collapsed onto the straw-dung matting of the shed floor. A downed cow.

A downed cow is no joke as you simply can't 'up' them again just like that, not even with a tractor, because they just go straight down again. And if they're in a shed, you can't haul them through the door either, as they become twice as wide when on the ground. If the worst happens and she dies, you will probably need to knock a hole in the wall to drag her out for removal.** Then you'll need a JCB with a very big bucket.

Ken was out, so we rang John, another dairying neighbour. Before Anne had even finished her sentence John

* This is not as outrageous as it sounds. Artificial insemination allows even the smallest breeder access to the finest bully fathers. The service is quick, cheap, and cuts down hugely on post-coital ennui.

** There is an alternative, which you don't want to even think about.

shouted, 'Milk fever … I'll come right round,' and banged the phone down.

In the panic of having to do all the spring preparation and sowing on her own Anne had forgotten about the theoretical risk of milk fever, and I was no use at all. So now Daisy was down. Coma and death would surely follow, unless …

I had visions of a scene from ER, with poor helpless Daisy lying prostrate on a stainless steel gurney, tongue dangling and drooling, while Anne held her hoof and gently patted it, and I panicked in a corner somewhere, next to a row of gleaming monitors going ping. John would feel her pulse and shout things like '13.2 per cent APR' and '78 rpm please', then '1500 C3PO please, Jules. Stand back everybody' and would whack a pair of dustbin lids onto poor Daisy's bosom. The lights across Dyfed would momentarily dim, while we all hold our breath. Again, '25,000, Jules. Stand back.' The lights dim to brown and the smell of singeing brisket fills the air. Then a little sigh … she twitched. Joyful exchange of glances. A hug or two. I slap the surgeon's back. 'Nice one, John.' He cranks himself slowly upright, sweat dripping from his furrowed brow. 'We were lucky. This time we were lucky. Half a junior aspirin twice a day.'

John roared down the drive and actually skidded to a halt outside the shed. His equipment amounted to a couple of pint bottles full of liquor, a length of rubber tubing, and a whopping great needle.

He rushed in, knelt down, felt for Daisy's jugular, and rammed the needle in. The calcium solution slowly drained from the bottle. I fetched a second bottle from the car. Anne fetched a cup of tea. Daisy was going to be alright.

Within ten minutes of the calcium hitting her bloodflow,

she began to get some colour back in her cheeks. Within twenty minutes, she was back on her feet and guzzling her sweeties. Utterly remarkable.

John explained that the problem was caused by the mother robbing her own bones of calcium in order to build up the bones of her calf. An extraordinarily touching gift. All Mums are susceptible. Even people: we call it hypocalcaemia. Cows of course are exceptionally susceptible as they have been bred down the centuries to produce lots of milk: 'Drink up your milk; it's good for your bones ...' we tell our kids. We don't think of the cow's.

* * *

The following year, John came round and gave her the Magic Calcium in advance, as a wise preventative. It worked. She just looked a bit pale for a while, but didn't succumb. She had her calf, the lovely April, and all was well.

As we now had our replacement for Daisy, we asked the nice AI man if he could give us a proper beefy calf next time. Perhaps an Aberdeen Angus? Certainly he could. Just hold still for a moment, modom. Think of the Empire, possibly? There. Cigarette?

Nine months later, in the springtime, John came round again and gave Daisy a prophylactic bottle or two. We were getting the hang of this ...

And so was Daisy. With no trouble at all she had her calf, a spry and jet black Angus bull, immediately christened Jimmy.

But the morning after, we noticed the wobbly signs of milk fever again. The prophylaxis had not worked after all.

Daisy was now at least ten years old, quite an age for a heavy milker, and Jimmy was probably her eighth calf. Being almost constantly in milk had taken its inevitable toll on her. She was just about worn out, poor gal.

John gave her another pint or two of calcium, but we could see it wasn't having the previous miracle effect. She did pull through, but it was a slow business. It took hours and not minutes, and it was a week before she looked more or less normal. Even then she was slow and dull. We didn't need an expert to tell us that she'd had her last calf for us.

But for the moment she was giving gallons of milk. Plenty for us and for Jimmy, and still some over. We bought in another Jersey bull calf from Nick and Sarah next door. As he was a Jersey tan all over, even under the arms, it was obvious he was going to be called 'Spot'. He helped Jimmy out with all that milk.

We over-wintered four cows. Old Daisy, Young April (herself in the family way), wee Jimmy, and daft Spot.

* * *

By the spring of '87 Jimmy was proving to be every bit as boisterous and stroppy as Ken had warned us he would be: 'Anguses are known for it. Don't turn your back on him.' And Spot was ganging up with him. It was only a matter of time before one of them caused an injury to one of us.

Time to go: one grey morning Ken came round with his trailer and we inveigled Jimmy and Spot into it.

Oh dear ... we knew they had to go, and we were quite sure they *would* go, but it was still hard. Wally had been hard, but we thought we'd get used to it. But we hadn't. 'You wanna come along for the ride?' Ken asked. Last time I'd refused. This time I couldn't.

We drove to Llanybydder in moderate silence. Ken had been sending animals for slaughter all his life, but he didn't make fun of the city softy sitting next to him. At the abattoir we drove into the pound and a man in a mottled apron and smeary wellies came out to meet us. Ken opened the trailer gates and between them they shooed the calves through a big door, into ...

I was silent all the way home. What were those two little beasts going to face? Fear, for sure. Strangers. The smell of blood. Noise, mechanical and animal. Slippery footing ...

* * *

By the end of March, April was showing signs of being ready. A week or two later she started pushing. It was her first effort, and not surprisingly, it took a long time. But eventually, there were the shiny little hooves! Then the bony knees! Then the snout! Then...

... then we realised the snout wasn't breathing.

The rest of the birth proceeded according to the book, but the little calf, another little Bambi, just wasn't present. No breath. No life. Perfect, but dead.

It's not uncommon, apparently, for a first calf to be still-born. Nobody seems to know why. Heigh ho ...

We dug a big hole in the orchard, in the next allocated spot between the trees, and hauled the newborn into the wheelbarrow. So *perfect* ...

A brief moment of Angelus, and that was that. Life goes on. We were getting used to death being a necessary part of life.

April came into milk with a vengeance, and Daisy was still producing, too. We bought in three calves to cope with the flood. First of all a Murray Grey-Jersey cross from Nick; then a Hereford-Friesian cross from the mart; and a month or two later, another Jersey bull: Dopey.

The feeding routine became a bit of a circus. We'd milk April out and keep what we needed, then split the rest between Murray (the Grey) and Panda (the black and white Friesian). They got a bucket each and a few sweeties as well.

We left Dopey on Daisy, as 'her' calf. However, Panda soon twigged on that Dopey could only suck one teat at a time, so she worked out a strategy of approaching Daisy from behind while Dopey was busy suckling from the side. It was a job disengaging her once she was locked on, as she was a fine big girl, with a suck that would strip the thread off a bolt.

One way or another everyone got enough and both cows were suitably relieved. Everyone was happy. Except possibly Daisy. She was giving signs of having just had enough of this mothering lark. We sometimes had to hold her still while Dopey (and Panda) fed.

In fact, the two cows were giving so much milk we even bought in a third calf from Nick and Sarah. Another little Jersey bull, Dippy.

It looked as though we would be over-wintering six cattle this year.

In fact it would be only five.

– 9 –
Rotavators

We used to have an allotment on the Trent floodplain. Solid clay, it was, and impossible to reduce to a tilth unless you had a lot of time and a big Moulinex. The best technique I discovered was to dig when possible, smack the puggy lump about with thc spade as I went, and not tread on it until surface dry, otherwise it compacted into wellie-shaped rockpools.

When dry enough to stand on, I sliced it up with an army entrenching tool into wedges the shape of badly cut toast, before garnishing it with seeds, preferably large ones that wouldn't disappear for ever under the slabs or down the cracks. Small stuff like carrots needed hammering into the surface with the back of a rake. Parsnips needed a chunk of tilth placing on top of every seed to stop it blowing away. Onion sets were easier, but if the land dried out too much in the summer you ran the risk of them turning turtle and disappearing down the cracks, supported by just a couple of roots.

One year we had a very late wet spring, and a neighbour who had less time than I did, but more money, suggested going halves on renting a rotavator. I wasn't very optimistic: clay; wet; slimy; cost ... but we had little option.

It was an attractive orange colour, and 'not too heavy'. That was the good news. The bad news was that it took

John* four hours to do half his allotment, and we never got round to mine at all. What an evil machine. 'Not too heavy' meant it had no mass to counteract the swirling forces of the blades. Consequently, instead of scything and chopping the sward, it ran off like a mad thing, skittering over the surface like a mutant supermarket trolley, with John hanging desperately on trying to stop it overturning and going for his legs whenever it barged into a tussock. His temper wasn't best helped by the steady drizzle.

But after half an hour he got the measure of the brute, and by leaning heavily on its horns, and pushing and heaving, with an occasional encouraging 'Ole!' from me, he eventually got it to apply its 2½ horsepower** more or less vertically. But then he hit another snag. Once the trollop had dug a bit of a notch, it wouldn't go forward. And after a minute or two it *couldn't* go forward because it was too involved in plunging ever deeper and deeper, like a dog seeking out a recently buried hamster, and you can't *push* a machine out of a hole, even if it's 'not too heavy', and the land is dry and tractable. Our land was wet, greasy and claggy, and John had balls of clay building up under his wellies, and his gloves offered less and less grip on the slimy handles with every passing minute.

After four hours he gave up. One half of his patch looked like the Camargue, and the other half like the Somme. Defeat.

---

* I feel I ought to apologise for the number of 'Johns' you'll come across in this book. We once did a count and found we knew eleven Johns, one of whom was a duck.

** 'Horsepower'? Where do they get these terms from? 'Weaselpower', more like.

But it was useful experience for life on the smallholding. I had learned that what a rotavator needs is (a) adequate power; (b) weight; and (c) *wheels*. With a competent 5-weasel engine driving a pair of grippy wheels, the driver is free to steer and guide, rather than chase and wrestle. Adequate weight ensures that the force from the cutters goes downwards into the land; and the wheels and weight combined mean much greater stability, and hence safety. A skittering gremlin is a menace.

Our ideal would have been the classic Howard Gem, but they were very scarce round here, and too dear, anyway. Instead, we settled for a tidy Howard 352 ('One careful old lady owner; full service record; go slower stripes'). It's smaller than the Gem, but sturdily built, and has served us well for some twenty years. A key tool.

If you've never used a proper rotavator you'll be surprised at how similar it is to driving a performance vehicle. You can't fight it, or you'll end up overturned and in tears. It's just too heavy to wrestle with, and too dangerous once you're out of complete control. The knack is to keep your eye on the road ahead, taking due note of any rocks, string, spades, humps, children, etc., in the way, and then guide the machine through, with a gentle touch and occasional lean on the handlebars. It's a job that requires a skill all of its own. The rotavator is the one tool I never let visitors use, unless just to 'have a go' on a straight and even patch.

I've only once lost control, trying to remedy a particularly bad bit of ploughing (my own proud work). I went into a deep furrow at too sharp an angle and there was no way I could hold it. Blades and petrol everywhere. None splashed onto the exhaust thankfully. The only way I could have put the flames out would have been with endless hurled handfuls of wet earth.

'Where's the rotavator?'

'I've buried it.'

The only other mishap I've had was when I was hauling hard back on the bars, against the cut of the blades, to subtly alter a line of approach. Slowly but surely, both rubber grips slid off the handles. The machine trundled relentlessly on, leaving me sat on my dignity on the floor, still clutching the grips. Oh, how we laughed.

* * *

Rotavators need treating with extreme respect. Someone local ended up in hospital with severe gashes to both her legs after she managed to tip the rotavator over and fell onto the blades while turning it round in a polytunnel.

* * *

Anne isn't the only one who has managed to run the Howard dry. Because the engine doesn't have a filling level marked on it, and has no dipstick,* I have no way of knowing the true oil level.

The engine manual is, needless to say, no use at all. It shows me an exploded view of the machine (not reassuring) with every part numbered in miniscule blotty type, but that's about it. As

* Am I the only person who finds dipsticks infuriating? Every one I've used has needed at least five measurements taking, and even then I'm rarely convinced that the smeary mess actually indicates any sort of true level. And you never know how much oil you are supposed to top up with, do you?

Man I am supposed to Just Know how to operate and dissect a small engine; like *Woman* is supposed to Just Know how to calm a fretful child, or cook. Or *Kids* are supposed to Just Know the value of democracy and why debt is stupid.

So I do my best: I tilt the thing backwards every now and then and squirt two fingers of oil from a Fairy Liquid bottle into the filler hole. I need to do this quite often as the motor has recently taken to smoking even more than usual.

Let's cut to the chase: I ran it dry. Embarrassment, gloom ...

Alun got us going again and explained that I should hold the machine dead level then fill it to overflowing. Pity the manual didn't think to mention that. Thereafter I have carried round a carrier bag with the rotavator. It contains the tube of oil, an old screwdriver for opening and closing the stupid little oil-hole screw-cover thing, an oily rag, a not-oily rag, and a spirit level. It has never run dry again.

* * *

That Howard is a wonderful machine. You can even attach a trailer and a small plough to it. We have never got round to using this facility, otherwise we could take a couple of months off in the winter to drive back to Nottingham to see old friends. One month there, one month back, ripping a neat channel into the M40 en route. Both ways.

Fun with slurry

John of the Magic Calcium rang me one day in March to say that he was clearing out his slurry pit, and would I like some of the muck?

To an impoverished organic smallholder, any offer of 'stuff', especially a load or two of top grade fertiliser, is pretty well irresistible. True, I was still unwell, but I was actually out of bed most of the time, and was clearly on the up. I was still going to bed at 9pm and sleeping badly, but I could undertake light duties.

What could be easier than to drive the tractor and trailer the mile or so to John's place, sit there while his JCB dumped a couple of loads of wallop into it, and then drive home again? Ideally, I could then aerate the muck (to allow it to compost) by firing it against the yard wall, as per normal, but that job could wait until tomorrow, if necessary. Just a simple, rewarding little trip …

Anne wasn't quite convinced, but not dead set against it. Let's do it, then.

So I fire up Fergie, hitch up the muck-spreader, and zoom off into the frosty morning. It is one of those sharp, clear mornings where the frost feathers the tops of fence posts and radiates clarity into the air. A happy man, wrapped carefully into a couple of jumpers and his favourite home-waterproofed coat sits at the wheel, enjoying the Spitfire roar of the perforated exhaust. My hat fits, my earflaps are down and firmly tied, and I've even remembered to wear my gloves.

The joy of being out in the world again! You truly don't know what you've got till you lose it.

A slurry pit is a universal feature round here. Under a rational system of farming (i.e. a system not entirely dictated by cut-throat competition), all farms would be mixed, thus allowing a farmer to use his own straw to bed his cattle on. The resulting mucky mix would rot down into excellent fertiliser.

Under our actual system, because wheat won't grow well in the west, and grass doesn't grow quite so prolifically in the east, the two branches of farming have drifted completely apart, like oil and water. The farmer of the east has too much straw, and, until banned by law, used to burn thousands of tons of it. The farmer of the west has too much cow muck. It won't burn, and can't be fed to livestock (although they have *tried*, believe it or not), so what are his options? He can (expensively) import straw, which he does to a degree, or he can use the liquid muck (slurry) in other ways, usually involving huge trailer-tanks which spray the stinking mess onto the grass. It's an inefficient method of using the fertility, but there you are. Progress.

Until the time for spraying arrives, the slurry is scraped out of the cow shed and into a huge 'lagoon': basically a pit. Unless the farmer is very rich, and can afford a lagoon-stirrer (I'm not kidding), the sloppy mess gradually solidifies, especially as any waste hay gets dumped into it, along with dead sheep (now illegal) and general bio-waste from round the place. The surface soon hardens off and begins to sprout a variety of grasses and wildflowers. Every year or so the local paper carries a story of a child who goes missing.

John used mainly chemical fertilisers and wouldn't miss a ton or two of bovine stodge; hence his generous offer to me.

We had a cup of tea with Sylvia before beginning. Outside the kitchen window was the inevitable black and white sheepdog, wandering about on the lawn, looking for something to round up and stare at. It spotted me and came across to the window. I'm a big dog fan and waved a cheery greeting. To my alarm, the brute bared its teeth at me. 'John, should that dog be off a chain? I mean, look at him.'

'Oh, it's alright' Sylvia explained. 'He's smiling.'

I then realised his tail was going like mad and his eyes were alive with fun. So I smiled back, and he smiled wider. 'Can he ride a bike?' I wondered.

John organises my trailer and his hired JCB into position and fires up. The big yellow bucket sinks its jaws into the grassy sludge and rips itself back out again, roaring like a brontosaurus – thwack – and dumps it into my trailer. The whole machine shakes with the impact. It sets off for another gulp, but suddenly stops.

Puncture.

Those tyres are monsters. No farmer can contemplate fixing them himself, so John rings the specialist who will be there 'as soon as possible'. Today? 'Bit tied up at the moment. See how it goes ...'

So there we have it. Maybe today; maybe tomorrow. Maybe not. It's very cold and there's no point hanging around. My tractor is, inevitably, behind the JCB and can't get out. I have to walk the mile or so home.

That little adventure put me back into bed for a fortnight. I had learned a useful lesson, however: that within reason, I could do anything, but might have to pay the price afterwards. After my second winter of collapse I was beginning to face the prospect of coping rather than recovering.

I wasn't alone. Daisy never fully recovered from the milk fever, either. Her arthritis seemed to get worse too, and she began to look like the 'old cow' I had once called her. She became thin and listless, and began to hobble. Her days were clearly numbered.

– 10 –
On Hats

Guests at The Palace and smallholders are both riddled with anxiety over what to wear. Morning suit? White tie? The Order of the Golden Banana?

Or wellies? Trench coat? And cheap anti-bramble gauntlets that waggle and slide about like off-the-peg armour?

The essential difference is that The Visitor is likely to suffer mere embarrassment if he wears a purple garter and suspender belt instead of a pink and yellow one, but a smallholder faces pneumonia if he gets his dress code wrong. The prospect of two or three hours on Baa Watch in horizontal midnight sleet concentrates the sartorial mind most wonderfully. Chic is out; practical is in, big time.

Your hat, for example: a soft brim is fine for keeping rain off your glasses, but hopeless in a bit of a storm. Even if it's tied under your chin with three strands of baler twine (don't laugh till you've tried it) it's a constant and dangerous nuisance slapping all over your face. Hats need careful thought.

I have a large selection I can call upon, of various efficiencies. The simplest is an elasticated and towel-edged sun visor, which keeps sun and sweat out of my eyes during occasional hot days on the field. It was once a tasteless fluorescent orange colour. Some years on, it is an attractive mottle of browns, tans and ochres, where fire and water have worked their alchemy. A priceless garment, but susceptible to wind.

At the other end of the spectrum is – or rather was, now that moth and rust have done for it – a genuine Texan stetson with beercan ringpulls hanging all round it on strings. Anne maintained that it was just silly, and if I must wear it, old-fashioned corks would be much more effective against flies than hi-tec ringpulls. I maintained that it was actually *not* silly, but a vital piece of equipment as it not only guarded me from debilitating heatstroke (a constant worry in Wales, as you might know), but the ringpulls also offered one hundred per cent protection against mobs of kittiwakes pulling my hair out. This argument cut no ice with Anne, but the fact is that whenever I wore it I was never once attacked by hordes of kittiwakes. I leave you to draw your own conclusions.

The other hats are more sensible. A cheap bush hat, which sags when wet, like a camouflaged Tiffany lampshade; and a sailor's hat, a souvenir of Holland, with which I once unwittingly startled a visitor one rainy day by offering her the choice of either a Welsh Woolly Hat or a Dutch Cap. I've been more careful since. It fits, so I wear it.

Then there is a lightweight khaki job with a flopsy peak and a neckflap that ties up over the hat. Very light; very warm. But also very smooth and shiny and blows off instantly. With the flap down, your neck keeps warm, but the peak is too soft to be any use in the sort of weather when you want the neckflap down. In other words, this very carefully designed hat, made from wonderfully efficient materials, is almost completely useless: because you can only actually wear it when it is warm and still: when you don't need it.

Then there is the Russian fur hat. Fantastically warm, but a bit tight, and makes my eyes bulge unattractively after a while. Also, no peak at all, which seems a bit pointless in a posh hat.

And the construction hat of bright yellow plastic, worn for chainsaw adventures in the dingly dell. Almost completely useless also, because if you trip the first thing you lose (after your balance) is your protective helmet.

But it was a kindly gift, and does keep the rain off your head ... by redirecting it straight down the back of your neck. 'Design', eh? But if you're daft enough to need protection from branches falling on your head you're not going to last long anyway. Crossing the street will do for you sooner or later. Or opening a jar of pickles.

So, after all these exotica, what is the favourite all-purpose titfer? It is, I'm afraid, the woolly hat. We have several, mainly homemade by Anne, of all sizes and thicknesses. No peaks, but they are warm, waterproof, light, easy, almost perfect. Made from the right yarn, and from a certain angle, they can also appear very fashionable, but hey, you can't have everything.

My ideal hat would be large, warm and waterproof. Something like a thick felted velvet cloche glued inside a WW2 tin helmet, only bigger, so the rain doesn't go down your neck, buckled firmly under the chin.

Or – just thinking aloud here – how about ... say ... *three* woolly hats (nested Russian-doll style), a small bin lid, and several yards of baler twine.

I've got most of those bits in the shed, come to think of it ...

Excuse me a minute.

* * *

The best hat for rotavating was that orange peak thing. By definition you would only be rotavating when the weather was warm and dry, so all you needed was something to keep the sun out of your eyes. Vital if you were not to chomp up an important crop by accident.

One crop I did chomp up a lot of was yarrow. My excuse is that I didn't realise it was a crop.

Yarrow is a white umbellifer of about dog height, famous as a wound-healer in the days of Achilles, and also used by the ancient Chinese for casting the I Ching. Perhaps most oddly, yarrow is the basis for Rudolf Steiner's biodynamic preparation No. 502: lots of yarrow flowers, sewn into a stag's bladder, hung up for the summer and then buried over winter. The resultant compost, Steiner says, is capable of revivifying even the most exhausted soils, even when used in miniscule quantities.

Silly superstitious nonsense? Ask any prosperous Dutch or German biodynamic farmer (there are even a few British ones) who has been using the system for years and knows that it works. My friend Henning tried it once as an experiment and was quite certain that his carrot crop was much better than normal as a result.

The 'problem' with biodynamics is that orthodox scientists don't have an explanation for how it could possibly work. Therefore they do the time-honoured, if alarmingly unscientific, thing of ignoring it, on the basis of 'It *can't* work; therefore it *doesn't* work' (as for dowsing). This attitude will change one day, but it won't be tomorrow, as the Materialist culture is incredibly deeply ingrained.*

* (SfaS): where I demonstrate as clearly as I can that Materialism is actually a wildly irrational philosophy and can thus not be logically sustained.

* * *

Meanwhile... yarrow...

We had a lot of it suddenly spring up where I had ploughed. I spent hours yanking out armfuls, wondering what we could do about it. What, indeed?

Working on the principle that the best way of getting rid of a problem is to sell it, I wondered who might like to buy it. This was before the days of the internet, and needed a long chain of letters, but eventually I did find a Swiss herbalist, working in London, who thought he could find a use for some nice dried yarrow. We stuffed two big paper sacks full and posted them off. They brought in the equivalent today (2006) of £80 or so. Very useful.

Obviously our next thought was to try other herbs. We made enquiries and tried our hand with a dozen more or less exotic plants. Most of them originated in North America and they all had odd folksy names. Pokeroot was one, Adam's Navel may have been another, and quite possibly Hagblotch. Don't recall.

But none of them really got the hang of it. Too wet, I expect. They all fizzled out, disappeared, or died.

One other exploration was doomed to failure from the start if I'd paused to think. I wrote to a homeopathic pharmaceutical company offering to grow herbs for them. They said no, but maybe that was no bad thing, because if we'd been successful they would have required smaller and smaller quantities every year.

A holiday and cheese

What do you call those things where you go away for a while and relax, and droop about and sleep a lot? Well, we thought we could all do with one, just as soon as we'd got the main crops up and running.

And where did you go, back in the pre-'90s, if you had very little money and wanted a holiday which would supply plenty of stimulation for the kids (Paddy was fifteen and Cait was eight) and plenty of chill out for the parents? Yes, an old-fashioned holiday camp.

We didn't quite trust the Reliant to get us to Minehead and back, even if our collective sanity could stand being cooped up in a second-hand space capsule for six hours at a noisy and bumpy, if recklessly steady, 30mph.

Anne was knackered, and I was still out to lunch most of the time, so driving wasn't an attractive option anyway. Let's have a trip on a train!

Between us we had a great time. Free fairground rides; free entertainment; free wandering about; and, as Paddy and I discovered, free snooker. We played a lot and came to know the various groynes and bunkers on the tables pretty well. Paddy won our own personal prize for the best break (twelve) and I was satisfied with the best pot: in off three cushions. It may or may not have been a fluke; opinion is still divided. My best break was four, and I was delighted with it.

Anne did a lot of dawdling and reading, and Cait tried everything on offer from trampolining to go-carts. We came home well-rested, and very grateful to the ticket-collector who took pity on us squatting on our suitcases in the corridor and let us into an empty first-class compartment. A great end to a great holiday.

* * *

Back to work: weeding, mainly. A week away means a lot more work to do when you get back, but oh yes, it was worth it. I'd pretty well shaken off the exhaustion and was ready to jump back in at the deep end.

And even dear old Daisy looked a bit livelier with the better weather.

It was about now that Anne began to experiment with cheese. We'd got the hang of butter (SfaS), but cheese was something else.

There are two basic types. Soft cheese and hard cheese. Soft is pretty easy; hard is pretty, well ... hard.

In principle, you can make simple soft cheese by just letting milk go off. You may not have much luck with ordinary shop milk though, because it has been pasteurised, which means that all the bacteria in it, good and bad, have been killed off.

Obviously we were going to use our own raw milk which still had all its microbes intact, and we were looking to cheese as a serious way of storing the energy in the milk. Anne needed to experiment.

She started by just letting raw milk go off, to see what it would be like. The first batch was pretty good, but the second one was horrible. There's lots of different bugs out there. How do you know how to get the right ones on your side?

So she read a few books and set to work, trying various 'starters', with and without rennet (extracted from a calf's stomach; no, honestly), or veggie rennet, and lemon juice. One way and another she ended up with the required curds and whey.

The curds are the important part: they get squeezed in a cloth to make a compact mass which eventually becomes the cheese. The whey becomes either a drink for yourself or the dog, or in our case, for the pigs. Or you just chuck it … awhey, preferably onto the compost.

If you fancy having a bash yourself, you could try this:

- Pour 2pts (1 litre) of whole (i.e. unskimmed) milk into a heavy-bottomed pan.
- Add 2 tablespoons of lemon juice, and stir.
- Warm gently and keep stirring until curds and whey separate (2–5min).
- Drain through muslin or an old sock (clean is best).
- Leave hanging over a bowl till the drips have stopped.
- Add salt to taste.
- Add a few chopped chives or spring onions or whatever you fancy. A handful of turnips, possibly, or a watermelon or two if feeling bold and rather silly.

This method won't give a lot of bulk, so why not try:

- Warm 2pts (1 litre) of whole milk into a heavy-bottomed pan, to blood-heat.
- Add a carton of live natural yoghurt. Stir.
- Cover and leave in a warm place to 'incubate' for 6–8 hours.
- Add 2 tablespoons of lemon juice. Stir.
- Proceed as above (… 'Warm gently', etc. …)

This technique gives more volume, finer texture, and more flavour. A light lemon taste often comes through, and suits it fine.

Add whatever you like to amend the flavour.

Experiment!

Anne experimented plenty, and was soon supplying us with a regular stream of quality soft cheeses, and some delicious yoghurts to boot.

Then came the real challenge: a hard cheese. For this you really must use the right gear: rennet (or veggie ditto), and then you must be prepared to spend hours poring over the semi-mystical procedures that aim to balance milk type with timing, heating, chopping and pressing. You might also add 'phases of the moons of Saturn', and 'colour of wallpaper', as the whole process is a pretty tricky one to get consistently right. The end result will definitely vary with what the cow has recently been eating, for example, so I'm sure you get the drift. Just compare a Caerphilly with a Danish Blue, or a Cheddar with a Camembert.

Anne soon realised that she needed a proper press to make a hard cheese. They cost far too much to buy so I made one out of an old pallet and a car jack. It's really just a double rectangular frame, about one foot by two, joined (or separated, depending on how you look at it) by a couple of cross-pieces at the top and bottom. The drained curds are pressed firmly into a mould (in our case, a cake tin with holes drilled in it) and placed on the bottom cross-piece. You then cover the cheese (still in its cloth) with a 'follower' – a disc of wood, just big enough to fit inside the diameter of the cake tin.

The fun bit comes next when you build up a suitable

number of blocks and shims and insert the car jack, so it presses up against the top cross-piece, and down against the follower. Slowly but surely, more whey than you ever thought possible drips out and all over the top of the table, then off the table and onto the dog's head.

So you balance the whole contraption onto a big roasting tray and slowly crank up the pressure again. More and more whey oozes out. More pressure, more whey.

If you don't start off with a big enough bulk, it is all too easy to make a hard cheese that resembles a small stale chapatti. Think big, is the rule. 'Really big', is a better rule.

But 'big' means more labour, and more time too, especially as you have to wait for months for the new cheese to mature enough for you to tell whether all that work has produced something edible (never mind titillating) or not. You can't rush cheese.

And there's the rub: despite a couple of moderate successes and a couple of moderate flops, we came to realise that cheese-making is an art that would require more time than we had available.

So now my lovely cheese press is relegated to the top of the cupboard for visitors to enquire about. I tell them it's a gibbet for hanging rats, two at a time.

We have friends who have persevered with cheese-making. They too found that it tended to become a full-time job. So much to learn and gauge and judge. So hard to get consistent results, and thus hard to find a regular market for their efforts.

We are now deeply respectful of anyone who has mastered the Art of the Cheese.

* * *

Time caught up with Daisy. As autumn approached it was hard to watch her struggling to get up off the grass. Arthritis and general weakness were upon her. It was time.

Ken took her for us. I guess after rearing half a dozen young of her own, and providing milk and butter and cheese for six of us for a couple of years, her final gift to the world would be to feed a hundred family cats and dogs for a week or two.

Thanks, Daisy.

– 11 –
Handtools

My friend John has a theory that power tools save time and effort. My theory is that they make work, headaches, and great big holes in your patience and piggybank, not to mention the palm of your hand if you're not paying full attention.

Heisenberg's Uncertainty Principle (Domestic Power Tools) states:

> *the only uncertainty lies in the precise timing of when a Power Tool will let you down, either catastrophically or merely malignantly. It is hundred per cent certain that it will do both, at some point; possibly concurrently.*

Consider: all things break; powered things break quicker and more expensively than unpowered things. And what happens when your SuperShiny Thing-O-Matic Deluxe disintegrates in a shower of green sparks and clotted grease, half-way through a tricky paint-stirring job? If you can fix it yourself (a remote possibility in this age of electronic chips and shattering plastic mouldings) it takes hours and is never quite right again. If it's beyond the succour of a big hammer or a little screwdriver, then you have to drive fifteen miles to a man who hasn't got quite the right part and come back tomorrow – no, better make that next Tuesday ... Come to think of it, don't bother till after Christmas. Meanwhile the job you bought the tool to do is not getting done, and the paint is developing a skin like a customer service manager.

I'm not a Luddite in this matter. I do believe in appropriate technology. I refuse to hand plough ever again, for example. But just because a machine is available, it doesn't mean you should use it.

* * *

The traditional garden spade and fork are overrated tools as well, even if singularly unpowered. My grandad used a spade with a blade the size of the *Daily Mirror*. You could bury a horse in one of his bastard trenches and his fork would have found favour in a Yankee whaler. Much too big! What's the point in lifting 30lb of soil once, when you can lift 10lb three times? You work to a better rhythm, and don't risk popping all your vertebrae.

And while we're on the subject: why is England the only country in the world that still uses that practical aid to a permanent rupture, the huge square-ended shovel? The French, Germans, Italians, and Americans don't use them. Neither do the Welsh or Irish. And why not? Because they are a very stupid design compared to the shovel with a long handle and a heart-shaped blade. I have two of these. A small one which is good for light, quick work; or digging, even. The bigger one is for shifting bulk material from A to B. They reach into corners; you can scrape with them, and lift them high overhead; you can mix cement with them, load the mixer, and clear ditches with them; you can 'chuck' with them. You can't do any of these things with the British Bulldog-style blunt instrument. Well, you can do some of them, but it's an awful strain on the back. If you haven't yet tried a proper heart-shovel, you're in for a treat. I bought my first one on holiday in Ireland. I saw it in a shop and thought, 'These are the people who shifted millions of tons of earth and rock, digging the English canals and navigations. What tool do they really *know* about ...?'

So, while John is looking down the back of the sofa for the scissors to hack open a bubble-pack, so he can put new batteries in his torch so he can look for his socks, so he can change his clothes, so he can go out in the car with his broken ElectroSuperTurboMatic Battery-Powered Spade, to the spares man whose lunch hour it invariably is ... I chip and chop dozily away and get the work done twice as fast.

Tip of the Week: if you rub your handtools down with old Mazola you will always bear with you the ethereal fragrance of chips.

* * *

Bread is something else Anne experimented with. We had long since given up on supermarket white, which is made entirely from recycled beer mats, and tastes of nothing, not even of ink or beer stains.

Proper baker's bread is better, but is still pretty insubstantial stuff. Too fluffy for our taste. What we like is bread that is nourishing, tasty, and al dente, without being dry and hard work. A slice of proper bread, when dropped, should land on the table with a dull thwack, and not collapse under the impact, like fresh shop bread, or bounce and rattle, like yesterday's shop bread.

The principle of bread-making is very simple. In fact, it's

the same principle used for beer- and cheese-making. You introduce a suitable microbe into a mass of raw material, and there you go. Everything else is a question of finesse.

For bread, you need baker's yeast, which works fast. The little yeasty people eat the floury stuff and produce indigestion, which makes the bubbles in the bread.*

The freshness of the flour, the grain it was made from, the temperature, humidity, the position of Venus relative to Aquarius ... they all affect how soon and well the dough rises, or 'proves'. Some recipes advise knocking the dough back and letting it rise a second time, which always seemed like too much work to us.

Anne eventually hit on a recipe that allowed her to cook eight delicious loaves at once in our Tirolia woodburner. We did have a failure or two en route, however. One loaf we could only cut with a saw. Crumbs everywhere, and a real effort to chew, despite all the butter and jam it would hold. But it's all part of the gig.

When we eventually got a microwave, Anne experimented some more. Now the microwave is the only way for us.

Her recipe, based on Doris Grant's classic (which is also the basis for Delia Smith's basic loaf) is as follows, if you'd care to have a bash:

To make four middle-sized loaves you need:

- A microwave; ours is a 700w model.
- 4 microwave-proof loaf 'tins'.

* You really don't want to know what yeasty people's by-product makes the alcohol in beer.

- 3lb (1.5kg) of strong (breadmaking), wholemeal, stoneground, organic flour.
- 2 sachets of 'instant' yeast (quicker than 'dried'; lasts longer than 'fresh').
- A heaped teaspoon of salt (or as you will).
- 2 pints of milk (or water or a mixture).

1. Mix dry ingredients thoroughly (we use a Kenwood).
2. Add most of the liquid. Mix, knead, thump, and pummel.
3. If needed, add the rest of the liquid: you are aiming for a dough that is 'just the dry side of sticky'. Good luck.
4. Divide dough into four, shape to fit, and charge your four 'tins'. (You might like to roll the 'loaves' in dry porridge oats (or pinhead oats, or sesame seeds, but not Smarties) after shaping, for decorative purposes.
5. Leave in a warm place until doubled in size.
6. Cook one loaf at a time for four minutes on high.
7. Remove the loaf from the 'tin' and cook for one minute on high, upside down.
8. Leave to cool if you really must.

Anne adds sunflower or pumpkin seeds to the mix. Obviously you can add whatever you like. Bits of olive, or sundried tomatoes, if it's been a hot summer; or sundried potatoes if it's been altogether *too* hot.

Three of our four loaves go into the freezer. But they don't stay there for long.

The microwave method is quick, easy, and perhaps surprisingly, much more ecological and fuel-efficient than using a normal oven. You need the power on for a *total* of 20 minutes, as opposed to 20 minutes just to warm up an electric oven, then another 40min for the actual cooking.

One caution: you won't get a crispy shrapnel crust on this bread – but that's no bad thing if you value your gums or your sight.

Visitors have always enjoyed this bread but feel faintly cheated when Anne explains it was made in a microwave. It's amazing what fantasies people carry round in their heads. We are smallholding country folk; therefore, they seem to think, we should bake our bread on flat stones in a peat bonfire, poking it with a sharpened antler, while screaming imprecations at the forces of darkness.

We have tried this method, obviously, but it's not as good as the microwave.

Ol' Paint

That diesel tractor, casually christened 'Ol' Paint', remained in the open shed as a constant caution to us. Whereas Fergie looked sprite and perky, Ol' Paint looked pretty much ready for the glue factory. I only once tried using it for proper work and just felt sad the whole time, as if I was asking too much of it. I knew I didn't have the resources to renovate it, so it was never going to get better. What to do?

The only thing: sell it.

We'd met a good few local smallholders over the previous five years and were constantly meeting people who had just started up and were looking for kit. I rang a few and one of them came to look at Ol' Paint.

'Is this it?'

'Yes.' (*Well of course it is. What did you think it was? A Euro-fighter?*)

'Bit oily.'

'Yes.' (*Ah, you spotted that. We're going to get on fine.*)

'How old is it?'

'Er ...' (*We've never asked it.*) '…'bout thirty years, as a guess.'

'Bit oily.'

'Yes … as I said on the phone, it needs a proper strip down and recon.'

'Does it work?'

'It seems to. All the tackle works, anyway. Hydraulics, brakes …' (… *although the mini-bar's getting a bit low.*)

'How much are you asking?'

'Like I said on the phone: £250' (… *being what we paid for it, actually.*)

'Yeah ... alright then.'

'Good.' (*WHAAAAAAA???? No haggling? No pointing out how oily it is? No laughing and sniggering even? Thigh-slapping, bent double in hysterics?*)

'I'm pretty good with motors. I reckon this one will come up a treat.'

* * *

Isn't human nature an endless surprise? When we first bought Ol' Paint, I was just plain pleased that I'd managed to dredge up enough concentration to achieve the purchase and then managed the drive home in the rain.

Then I was dispirited and depressed when I realised I'd

bought what the trade calls 'a nail', and wasted £250 precious £s on something (a) useless and (b) unsellable.

Then I was overjoyed that we'd just been offered our money back, and had even made a small profit, if you count the cultivator that came with it.

But then … when it looked as though the purchaser was going to convert the wreck into a thing of joy and much greater value, *Envy* surfaced …

Obviously, I whacked it on the nose with the starting handle and it retreated into its lair again, but not before its presence had been shamefully noted.

Quick handshake. Quick exchange of a bundle of grubby notes and that was that. No receipt. No log book. No insurance. He just poked at it, prodded it, waited for the smoke to disperse a bit, and drove off. Never saw him again. Or Ol' Paint.

I sometimes wonder what became of the poor old beast.

And Ol' Paint.

– 12 –
Hello Cheeky! (and Farewell)

I once met a professional shepherd who regarded his sheep as his family of somewhat eccentric but wonderful children. He spoke of them in the way Jane Goodall speaks of her chimps.

His tales of individuation made me look more closely at our own flock, and the closer I looked, the more individual each sheep became: more like a dog than an anonymous item in a group. And once you have recognised a personality, it becomes hard to see the animal as mere meat in the making.

Cheeky, for example: she and Star were our two first sheep. Star died almost immediately in childbirth which was a sobering introduction, but Cheeky lived on with us for fifteen years, latterly as a pet-pensioner. She was a pure Jacob, a so-called primitive breed, possibly from Palestine. Certainly, she had an 'unimproved' look to her: small, slight, horned, and with a capricorn glint in her eye. And the famous brown and white patched coat, inclined to armpit kempiness.

Other sheep came and went for a variety of reasons, but Cheeky stayed on as the undisputed Leader of the Pack. She was the one who would first sense your approach, long before her dozier sisters. The others would eye you with either a startled stare or a casual glance, but Cheeky held you in cool appraisal. The others looked away; Cheeky looked on.

What for? Who knows, except that she knew people

mean change, and if there was a bucket of oats in the offing she wanted to know about it. Likewise, if there was to be a general roundup, for whatever reason,* she needed to look for the best escape route. Where Cheeky led, others would follow, but as she was so much smaller than the tubby Southdowns or the lankily aristocratic Leicester Longwools, she often left them behind when she squeezed through the tiniest gaps in our defences. She never actually ran away though, unlike some of her more volatile companions.

She was an excellent mother. Each year she was tupped** by Ken's great Roman-nosed donkey of a Suffolk ram, and gave birth, very easily, to a singleton, or occasional twins. She kept a good eye on them, and only showed anxiety if they became too familiar with The Management. One of her lambs, a jet black beauty called, obviously, 'Hole', was extraordinarily chummy until we treated her for a painful patch of flystrike … then she never trusted us again. I'll swear I heard her Mum bleat, 'Yah-hah! Told you so!'

Cheeky, of course, didn't catch flystrike or foot rot or any of the other ailments over-bred sheep are prey to.

One year her own baby was late arriving, so she lambnapped a little Longwool and wouldn't give it back. We had to forcibly separate them, not just from spite, but because she would need her colostrum for her own late arrival.

However, her own lamb was still-born, which came as a great shock to Cheeky, the true professional. She wouldn't let

* … like dipping, foot trimming, dagging, worming, perming, shearing, or, most romantic of all, examining round the bottom for maggots …

** A word so much more romantic than 'inseminated' or 'bonked', I think.

us 'disappear' it, and stamped and butted us away, as she would with any dog. For two days she returned at intervals of an hour or so to the lifeless bundle, nosing it and mumbling. No fox would dare intrude. Slowly she accepted the truth, and we removed her sad reminder.

She became thin in her last months, probably due to lack of teeth, and sat down a lot, and seemed to welcome some personal attention from us. Her wool dragged off in strips, and alarmed us once when we mistook a billow of fleece for her dead body. But no – she was still sat there, under a tree, gazing across the valley.

As we watched her fade, other memories surfaced …

… of how we'd once lent her to Ken as wet nurse to one of his orphan lambs, to whom she took an immediate dislike, and refused to allow to suckle;

… and of how she would walk quietly on a lead (unlike some I could name) when it was dipping time;

… and of her small lumpy horns that regularly got caught up in pigmesh fencing, until they snapped off, luckily to no ill-effect.

Then, one autumn morning, she was sat under her favourite tree; chin down; gone. We had a bit of a drip, patted her head, buried her in the orchard, then had another bit of a drip. The eternal mystery.

We remember her by happy memories, a homespun and knitted blanket on the back of the sofa, made from her soft fleece, a few photos, and a bit of broken horn in the drawer we reserve exclusively for string and whistles, foreign coins, old specs, broken torches, dead biros, and discredited credit cards.

Just a sheep. The sort you might be having for dinner tonight.

* * *

We were getting used to the idea of home-grown meat by now, but the actual fact of killing never became any easier. Our sheep and cows were definite individuals, not industrial units. Just think of how you might feel, sending your dog for slaughter. But it had to be done, we thought, so we gritted our teeth and got on with it. After all, what was the alternative? As far as we could see, you could be either organic or vegetarian, but not both (SfaS).

Little Red Racer and the Bismarck

The crops performed as ever, which is to say that some thrived and some didn't. The beetroot didn't do very well for some unfathomable reason, but the sweetcorn was spectacular. Because it's such a dodgy crop normally, we always over-sowed by 50 per cent and hoped for the best. This year, every plant just went for it, and each produced a couple of full cobs. We stuck a great pile in the freezer, but still had too many. Then we sold a lot at the Women's Institute market and to pubs and hotels in the district ... still too many. Eventually I found myself squatting outside the Bunch of Grapes like a Levantine trader with three trays piled high with sweet and succulent maize under a damp cloth. It was hot, and the village was filled with tourists. An occasional cry of 'Salaam effendi' caught the visitors' attention, along with a naughty little belly dance, and I sold the lot.

Not all the takings found their way home, alas. Well, it *was* hot, and the Bunch of Grapes served a nice pint.

There really is no way of knowing in advance which crops will succeed and which will fail. The trick is to remember that you can never dominate Nature and must learn to collaborate as much as possible, and to never expect to win them all. If it's a rotten bean year, it will probably be very good for spuds or onions, and vice versa. And of course, the less greedy you are, the more adequately will your needs be satisfied.

It is a salutary thought that if Western Man's endlessly increasing consumerist-wealth has been made at the cost of exhausting his soil, he will not be pleased with the final account: the Law of Cause and Effect will run inexorably.

* * *

The Little Red Racer (the Reliant) was no longer adequate for us. We needed to carry a dozen or two heavy crates of courgettes twenty miles to the wholesaler's, twice a week, and the LRR simply wasn't up to it. We needed a van. But we also needed something to carry the four of us about from time to time. Hence, we needed a van with four seats. Or something. What should we do? We should talk to Alun, is what.

It so happened that the Reliant's back brakes needed adjusting so I drove round to see him. While he was battering away at the rusted-up mechanism I raised the issue of a new vehicle. Could he find us a good second-hand van? 'Told you before,' he called from under the car, 'vans are either good or second-hand. Well-known fact. What do you want it for?'

I explained about the dilemma between load-carrying and people-carrying.

'What are you leaning on?'

'Eh?'

'Behind you ...'

'Oh … dunno. A limo. Some sort ... is the paint wet? Sorry.'

Alun slid his trolley back out from under the Reliant. 'That's what you want, man.'

I laughed.

'No, I'm serious.'

I turned and looked at the monster.

'Volvo estate. It'll carry anything "to the moon and back" like they say in the adverts.'

Was he joking? Surely …

'Good nick. Guaranteed 180,000 on the clock.'

I laughed again.

'What are you laughing at? I'm serious, man. "To the moon and back". The motor's good for 300,000 at least. In Australia they never wear out at all because the bodywork never rusts.'

Great Scott ... he *was* serious.

This car is *huge* ... and it's got a radio. *Four* wheels ...

'Talk it over with Anne.'

'Ah ... yes, but ... Nice idea mate, but right out of our league.'

'£850. How much would a van cost?'

A lot of money, but a lot of car. Alun swore it would be reliable and that he would take care of it. It would certainly do what we needed, and amazingly, the fuel costs wouldn't be nearly as much as I'd expected. And anyway ... how much *would* a van cost? ... a dodgy, second-hand, van?

Anne and I were taken aback at this radical option. Then we thought it over for a third time and could see no alternative. We did the sensible thing and said, 'Yes please.'

As we cautiously drove away, Alun called after us, 'Just don't use third gear.' Mechanic's humour.

In fact, I used third gear a lot for the first week of ownership. The car went so fast and smoothly in third that I frequently forgot to change up.

And it was so *big* ...

We often park our cars just outside the gate, and approach them from behind. On two occasions in the first week I rounded the back of the car, took a suitable number of steps and opened the door to drive off. But it was the back door I'd opened. That's how much bigger than the LRR the Volvo was. A whole door's-worth.

We christened it The Bismarck and began to fall in love with it.

* * *

Illness knocked at our door twice that autumn. One night Paddy developed a splitting headache and couldn't move his head properly. Paracetamol didn't help. The doctor came out and rang the hospital. Meningitis it was. They gave him an epidural (I can still hear it) and laid him up for a week. He came out cured but groggy and with a boxful of antibiotics for company.

A week later the ME returned ... again. Two months earlier, this time.

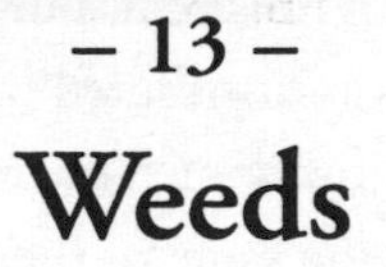

– 13 –
Weeds

I do feel sorry for weeds. For them, success means disaster.

By definition, a weed is a highly successful plant which just happens not to agree with your idea of where it should be. It is also a plant which is so spectacularly adept at surviving in the great botanical rat race out there* that it can germinate, sprout, grow, blossom, and seed faster than its competitors. Success!

Unfortunately, its competitors have as their ally the most intelligent, resourceful, and ruthless animal on the planet, who is determined to ensure the survival of many of the more effete and soppy species for its own gastronomic and sybaritic use. Therefore, the more adapted and 'successful' species are doomed to be radically removed and cast into the outer darkness of the compost heap or stuffed into a cow. Disaster!

Every plant-loving gardener massacres plants by the tens of thousands every year. A market gardener kills millions every week. And their only crime is that they are seeking their place in the sun. Tragic, isn't it? For them, natural selection has been superseded by the whim of a bloke with a sharp stick.

I find it does help to bear all this in mind when I go weeding. I know my victims are living creatures, not mere inert nuisances. So I can at least admire them as I mow them

---

* ... if you can have such a thing as a botanical rat. No doubt *GeneticNightmares-R-Us* are working on it even as I write.

down. Small consolation for them, but perhaps better than a casual dose of nerve toxin distributed by a man in a spacesuit from the inside of a multi-boomed tractor. However, common sense insists that it's all the same to the withering weed at the end of the day, if not to the creature whose place in the Great Pecking Order is to eat it.

Within limits, I actually enjoy weeding. Of course, enjoyment varies with the where, when and what. Hauling endless yards of case-hardened couch grass cabling out of the garlic beds in that terminally wet 'summer' of 1985 was rather less fun than gently knocking back a dusting of assorted weedlings from the bases of a thousand lusty courgettes, with light breeze, mild sun, birdsong, and not too many military jets splintering the air. What could actually be nicer on such a day? Broiling on a beach, getting sand in your wotsit, and finally returning home as Fermier Tartare, too tender to sit back in your seat? No, not for me.

Perhaps the worst weeding job ever was in our first full season when we actually forgot about one crop completely for several weeks.

We had this great big patch of ploughed and levelled soil, which Ken next door had taken the trouble to prepare for us, to get us started, and we just couldn't countenance leaving a single foot of it to revert to grass again. So we bunged in whatever seed we could find and hoped for the best. We tried to keep up with the weeding, and managed pretty well, except for the shallots. They gradually sank into the returning grassland and literally disappeared. Finally we did remember them, but then couldn't find them. They'd been planted with great precision via a prototype for my Patent Dibbing Tool (see

Chapter 30) and were rediscovered by precision, literally with a ruler. Each was then brutally weeded with a sharp hoe and a small spade. It took hours and was grossly non-cost-efficient, but what the hell. You can't let your friends down.

* * *

I'm constantly struck by man's ingenuity in his battle against the weeds. A farmer friend, who employed casual labour to weed his fields, designed and made his own hoes. The crops were grown on the tops of ridges (better depth of soil; better drainage). The weeds were on the sides of the ridges, and an ordinary hoe would degrade the ridges too much. The solution was a simple loop of fencing wire, pushed roughly into a coat-hanger shape, with the ends jammed into six inches of polythene piping. They worked a treat. Not kind to your back after a couple of hours, but that was not the farmer's worry. (Casual labour can be treated pretty casually, even by organic growers.) One inspired labourer brought along a drilled-out broom handle one day and stuffed his wire loop into that. Voilà! End of backache.

It seems to me that the right attitude and a bit of ingenuity and forethought can turn even the mind- and fundament-numbing job of cleaning up hundreds of yards of carrot row into a personally rewarding achievement.

If you regard the job as a chore, you are depressed and exhausted before you start. But if you regard weeds as worthy adversaries, you are halfway to success: your own dexterity and patience pitted against their sheer numbers and defensive ploys. Your carefully selected implement against their hypodermic pricks and daggers. Think: 'a thousand feet of beautiful carrots' rather than 'hours of backbreaking agony'. And while you're at it, you could do a lot worse than to take Virgil's lightly paraphrased but timeless advice: 'Why not go forward, singing all the way?' A happy weeder is an effective weeder, as I'm sure Virgil would have added if he'd had the time.

* * *

On the strictly practical level, you need two classes of tool for weeding: between-bed and in-bed.

The rotavator is the obvious between-bed tool, as it will uncomplainingly drag me round the field for hours, scuffling up weeds for little apparent effort. But it's noisy and smelly, uses ungreen and expensive petrol, and requires my constant attention so I don't inadvertently mince my way through the newly planted marrows, like some sort of demented Fifth Horseman of the Apocalypse (Motorised Division). Also, the rotary motion does not kill all the weeds. In fact most are merely half-buried, and will regrow in days.

A much better tool, in my view, is a splendid old 1920s wheel-hoe we bought at a sale for a fiver. This is essentially a pair of push-along handles attached to a twelve-inch wheel. Just behind the wheel is a jig on which you can bolt a variety of small field tools, even a mini-ridger. The swept-back weeding blades are best. You can get a fantastic leverage on them and even slice through big docks with a good hard shove. It will lift turf at a pinch. Once up the hill and once down, it takes out every weed between the beds. I like to think of it as Brazil waxing for gardeners.

Also, you can nip in between the courgettes with it and under the leaves (which you can't with the rotavator) and get right up close to the carrot beds (which you can't with the rotavator). And it doesn't stink or roar, or run out of fuel at the top of the hill. *I* might, but it doesn't.

It is silent and absolutely cost-effective, and will let me reverse and wriggle into tiny gaps if I want to … and as it slices weeds off at ground level, they stay sliced, and won't regrow. Even docks give up after a few slicings, as they gradually use up all the energy stored in their whopping root.

And, to my amazement, this apparently primitive tool is *quicker* than the rotavator.

What's more, it takes less energy on my part as well. OK, the rotavator does haul itself along, but you have to direct it and keep it balanced and you use a lot of nervous energy stopping it bouncing on a rock and swathing through the baby beetroots. The wheel-hoe just needs a rhythmic push-pull rocking motion which feels much less tiring. Better for the biceps too.

And it never needs servicing, topping up, coaxing, poking or even whacking repeatedly with a large spanner. It *never* breaks down.

And after an over-busy day, you can just leave it where it falls without worrying about it getting soaked and ruined – although, in fairness, you might *lose* it, as I once did, if the grass on the headland is a little over-lush ... round the shallots, say.

Twelve out of ten.

* * *

In-bed weeding requires more delicate implements. Our selection includes a miniature onion hoe, two sheath knives, an old carver with a snapped blade, a green thing with a sharp bit and a pointy bit, a Victorian table fork, and a queer device made by Wolf that looks like Captain Hook's kung fu prosthesis. It is of a general square-sided-question-mark appearance, the short side being sharp and the long side deeply serrated. General Patton would have worn one on each hip if they'd been around at the time.

All these devices have their particular uses. The kitchen fork is great for teasing round deep-rooted weeds that are growing too close to the crop plant to simply haul out. The onion hoe is good for scraping out tiny weeds round ... oh, onions, for example. And so on ...

My favourite is the sheath knife with the home-made wooden handle. You shove the tip into the soil by the base of the weed and tug and slice. Simple and effective.

Anne's favourite is the Captain Hook job: much more brutal.*

But, despite all the technology, about 20 per cent of weeding comes down to forefinger and thumb. And you can't do it in big floppy gardening gloves. I did once try using close-fitting ladies' cotton jobs, but they quickly became wet and slimy and not only useless for weeding, but quite unsuitable for the Lord Mayor's reception that very evening.

As a rough guide, using a combination of wheel-hoe for the rough, and an ergonomic hand hoe for the fine, one can keep an acre of veg effectively weed-free for a few hours' work a week.

---

* Just for the record: the most useless weeding tool I've ever tried is the traditional 'hand-weeding' fork. Also awful for weeding are the Dutch hoe and the draw hoe which waste far too much energy. A hoe should slice weeds off at ground level with a push-pull action. I favour the ergonomic Wilkinson's Swoe. There are others.

But a good attitude is what ultimately makes for effective weeding.

* * *

The bonus you get from a good attitude is that you get to really enjoy just being there. You're about as close to Nature as you can get (much closer than hiding behind the White Hunter in a safari park), and you fall to contemplating the mysteries of ecology and the fact that plants grow at all: a juicy vitamin-packed carrot from a dried up little dot of a seed? Oh, come on!! Yet it happens; by the billion; every day; everywhere. If it wasn't so commonplace, every journal in the world would be headlining this astonishing transformation as a miracle (SfaS).

Tiny Little 'Seed' Produces Enormous, Nutritious, and *Self-Replicating* Plant!!

Pope Alerted

And the lovely flowers on the 'weeds': the scarlet of the pimpernels; the mauves of the nigella; the forget-me-not blues. All doomed to be rooted out, with just a shadow of the feeling when sending off another cow or sheep. Again, you might think this pretentious. But it's the truth. If you've done it, you'll understand.

* * *

Having said all that ...

... the most efficient way of clearing a million or so weeds at once is to use ...

fire

... Not the traditional 'slash and burn' sort of bonfire favoured by our jungle ancestors and modern East Anglian prairie-makers, but a small precision flame. The principle is that if you can manage to boil the water in the leaf cells of a just-emerged weed, you stop it photosynthesising and thus kill it stone dead, leaving the ground clear for the crop, which, by definition, will be slower to emerge than the weeds. Clever stuff. A small, hot, precise flame is all you need.

We knew that bigger local carrot growers were experimenting with various Heath Robinson rigs carrying banks of propane bottles behind their tractors, incinerating everything on the surface of their seed beds and way beyond. They left behind an aroma of over-roasted coffee, smoking sprout-stumps, and melted pebbles. 'Try turning it down a bit more ...'

We were interested because although hand-weeding carrots can indeed be a sort of joyful meditation, it also takes a very long time when time is of the essence. A fast-weed option would be very handy.

We didn't want a cumbersome tractor-mounted machine, though. We just needed something portable; light; easy ... that sort of thing ...

The farmer's co-ops couldn't help ('Flame-weeding? What's wrong with nerve gas?'), except for some wimpy looking paraffin items that seemed to need a gallon of juice per row, and our organic colleagues all rated propane as a greener option anyway.

Surprisingly, a local gas dealer had the right sort of gear

(Danish, not British, I was sorry to note)* and made us up the rig we needed. It comprised a small bottle of propane, weighing just under forty tons at first impression, a pressure regulator, a length of rubber piping, and a wand with a control trigger and two three-inch burners at the end.

You need a dry day for flaming (or you waste most of the expensive propane boiling rain water up) and one with no wind, or you run the risk of setting fire to yourself and your poor innocent livestock; AND you use too much energy as the wind swirls your flames all over the place.

Impromptu Ox-Roast Lands Pyromaniac Prat in Hot Water with RSPCA.

'A Few Charred Bones' Pathologist Reports ...

The signal to burn is when you see the very first definite carrotlets amidst the bustle of chickweedlings, mini-groundsels, bits of bittercress and resurgent docks and thistles. There they are, just a couple of tiny carrots. So here we go ...

Given that it's a two-person job and that one person has the delicate and rather enjoyably skilful job of wafting the blazing wand gracefully and steadily from side to side across the bed, and that the other person has the job of humping a forty-ton tub of gas uphill, guess which job I ended up with?

Actually, it wasn't forty tons at all, but it was too heavy just to tuck under my arm. The technique is to hump-and-

* We seem to have lost the urge to *real* horticultural innovation to the same degree that we have swallowed aggro-chemical farming. And anyway, we are a pretty conservative society overall. 'How many Oxford dons does it take to change a light bulb?' 'Change?'

dump it forward a yard at a time, trying to avoid the wafting fairy's delicate slipper as you do so.

Progress up the field is remarkably quick in fact, at about the pace of a slow walk. But humping a heavy steel tub of explosives uphill at even a slow walk is not something you want to be doing for hours on end.

I did try to devise a trolley for the tub but it would need to fit into the path between two beds, which was only a foot wide. No chance.

In an ideal world you would till your seedbed and flame it after a few days to do for the first flush of weeds; then sow your carrots and flame again after ten days. But we never had enough time for that, and just the one flaming usually worked

A sort of apology

I do realise that all this is probably far more than you ever wanted to know about weeds, but please realise that this sort of stuff is what most of our summers were about. Routine, routine, just like any other job. True, we did have the advantage of being outdoors a lot of the time, and of being able to pick and choose our daily jobs within reason, but every farmer is a slave to the seasons. Some jobs just must be done, and miles of weeding was one of the most important. Hence our obsession with it.

Our aim was to strive for the maximum efficiency in every detail in order to prove that we really could earn a living from a measly five acres when conventional farming wisdom said you needed a minimum of forty.

a treat – except for the year when we miscalculated, and flamed too early. By the time we'd realised, the carrots were up everywhere – and so were all the, pardon me, flaming weeds, which smothered them a treat. Back to Captain Hook.

Regrouping

As ever, it had been a 'mixed' year, with a few alarums but plenty of positives: lots of what we like to think of as 'learning opportunities'. But as winter drew closer, we found we were looking forward to it more than usual. Time to recover, recollect, and regroup.

The big pluses were that Paddy had recovered, we had a shiny new house-cow and car, we'd survived another year's weather, and had increased our income a little. That was enough; something of a triumph, in fact. As for my own illness … well, we were getting used to it. We'd cope.

– 14 –
On Clothes

You can always tell farmers from smallholders at an agricultural show. The farmers all look like fashion plates from *Hoof and Horn Monthly* in their Barbours, proper trousers, and checked tweed caps. The smallholders all look like tramps who've let themselves go.

In Wales, every farmer seems to wear a snappy brown check jacket in laser-sharp Terylene, rather than the Barbour. The tweed cap is often worn heaved over one eye at a dangerously rakish angle.

Smallholders seem not to wear tweed caps. They wear woolly hats or weird ex-army fantasies, or even Drizabones and deer-stalkers; but never tweed caps. So it seems.

Nor do they wear Barbours, partly because they are not cheap you know, and partly because smallholders don't have any interest in fitting into a dress code of any sort. Most farmers seem to need to belong, somehow. More conservative, perhaps? Smallholders wing it.

So what does The Smallholder wear, if not a fashion statement? Boilersuits are well-favoured, largely because they cover your kidneys and grommits and can't pull apart when you bend over. Farmers use these a lot too, actually, but never go to shows in them.

Boilersuits are also easy to find on the bedroom floor in the morning, and 'one-piece' means you can't lose the other piece. But they can be a pain to get out of after a hard day's shovelling or fencing when your muscles have turned to jelly. One can so easily end up barging round the kitchen like a broken marionette, trapped limbs flailing wildly, scattering kettles and crockery off the stove and draining-board. I've been known to shout for help.

Comfort must always be the key, so thick woollen shirts score well, as do padded cotton ones, preferably in hysterical Canadian lumberjack checks (the McIroquois tartan, ye ken). Worn over a T-shirt, you can stay warm and even look just a little bit butch, without a jacket even in a light frost.

On top of the shirt, The Smallholder wears a warm but loose-fitting number from Oxfam; his partner buys hers from Oxfemme, of course. It might be a jerkin or a cagoule, or a parka of some sort. Whatever it is, you can guarantee it will no longer be within earshot of the catcalls of Fashion. Alternatively, the jacket (jackette) will bear the black, gold and red swoosh of the Wehrmacht. Army gear is understandably very popular as it has been designed to work in conditions almost as demanding as a Welsh summer.

Views are mixed concerning waterproofs. Most of us go through a phase of trying several different approaches, ranging from a cycle cape, through a two-piece suit of rubberised fabric, to a waxed jacket (*looks* like a Barbour; much cheaper; falls apart; burns well) and all manner of pick'n'mix outrages. We each decide which solution is the least worst for us, because, at the end of the day, rain means you get wet. That's not meant to be trite. It's just a fact, because if you're working, in rain or otherwise, you sweat; and if the sweat can't escape because you're cocooned in a plastic bag, you get wet from the inside rather than the outside. So take your pick.

Some people decide to accept Reality for once, and just wear something comfortable until saturated, then regard that as a free wash, and go and hang it in a barn to dry, then change into another ragbag of cast-offs which are relatively drier. Most of us know that wool knits into a wonder fabric that actually expands when damp, thus becoming rather more waterproof, and which also, astonishingly, physically gives off heat. Good enough for sheep; good enough for me, mate.*

I did once make an effort to design the ideal wet-weather coat. There was a market stall in Newcastle Emlyn selling ex-Deutscheflotte knee-length coats. Wonderful quality, with detachable linings, hood, pockets. Everything bar a built-in hammock and phone. But it was faced with cotton. Nothing wrong with cotton, but, unproofed, it does get wet very quickly.

The solution was Mesowax. Anyone who's ever owned one of those antique sailcloth and oak tents will know about Mesowax. You paint it onto fabric and it makes it (a) waterproof, (b) stiff as a board, and (c) twice as heavy. Wonderful stuff.

Picture the scene, one hot and steamy July morning, as Andy the postman squeals to a halt (yes, he is in a van) and sees before him a stout and bearded man wearing a thick navy overcoat, with the hood up, arms akimbo, like an overstuffed scarecrow, while his good lady wife is slashing about with a four-inch brush, enthusiastically basting him with what appears to be curdled custard.

'Bore da!'

* But why don't sheep all die of heat-stroke after a summer cloudburst?

'Bore da.'

'Feeling the cold, are we?'

'Go away and get on with your work, Andy Post, or I'll report you, and then you'll be fired, and your wife and children will starve, see.'

'Ha ha! Wela'i di, te, nutter saes.'*

Yes, well ...

The waterproofing process worked remarkably well, in fact, and I still have the coat. It's rather oddly striped and smeared for formal wear, I agree, but very useful in a storm. And plenty good enough for an agricultural show ... along with a busby and flippers, perhaps.

* * *

We went to a couple of local shows, but never really got involved. For one thing we couldn't afford the entrance fees on our ultra-low income. For another, we didn't have anything to exhibit in the way of prize livestock. Our little gang of runts and hobbledehoys would be a laughing-stock. And, at the end of the day, we didn't really see the point of entering the competition for 'six nice potatoes' or 'four identical runner beans', even if we'd had the time to spend on such fripperies. All our efforts had to go into earning a living. And anyway, there wasn't a category for the 'most exquisite garlic'.

We enjoyed seeing all the delicious cakes and jams and

---

* saes = saesneg = Saxon = English person (cf. 'Sassenach' in Scots). A term (usually) of amused incredulity rather than actual abuse. Oddly, though, neither Anne nor I can claim to be real 'saes'. According to our family surnames, I appear to be of Moroccan–Irish descent, and Anne is Cornish–Norwegian. Pure-bred mongrels, both of us.

children's paintings and so forth, but couldn't drum up much interest in things like the 'longest parsnip'. The winner (every year, same bloke) use to grow them in drainpipes and feed them on – well, I've no idea. Guinness and peanuts was suggested. His monsters were several *feet* long.

It's nice to wander round a show, once every year or so, and look at all those immaculately groomed sheep and cattle. I think Ken and Doreen have the right attitude to shows: 'Bit of fun, isn't it?', but some people clearly go at it hammer and tongs (well, 'tongs', anyway; see below). You can see their point, too. Farming is a business, and a show is a rare shop window for local stock-breeders. Ken's 'bit of fun' Border Leicesters regularly win prizes, and prizes affect prices. A champion ram will be worth hundreds of pounds or more, and so will his progeny. Farmers will travel the length of the island to buy good stock to beef up their own flock,* so it's worth taking a bit of extra time over presentation. Contenders will have their coat shampooed and endlessly brushed, and their toenails filed tidy and polished. Curlers are a possibility, and even a little light cosmetic assistance, believe it or not. Ken told us of when he first thought he'd better keep up with the trend and try the cosmetics approach himself. He went into the local Boots and started poking about in the racks. An assistant asked him if she could help. Ken said he was looking for some eyeliner but wasn't sure what colour and so forth.

'What sort does your wife normally have?'

'Oh, it's not for my wife.'

(*A nearby group of tourists show interest.*)

* You can't say 'mutton up', can you? Odd, that.

'Oh, I see, sir ...'

'No, it's for a sheep.'

(*Tourists exchange alarmed glances and suppress giggles and bleats.*)

I wonder how far this trend for tarting up will go?

Four-inch Heels 'A Step Too Far' Say Show Judges.

Only Flatties From Now On, Farmers Warned.

Post-Show Disco Pix pp. 12–13

* * *

Something much more up our path were farm sales.

A farm sale is a sort of one-man agricultural boot sale, and a very useful source of small stuff for a smallholder. And of course, a sale is somewhere you are bound to meet everyone you know to catch up on news, and exchange ideas.

But at a deeper level a farm sale is a serious social event, as it often marks the death of a man who's been farming the same land all his life and who has no one left to pass it on to. This might be because he has no heir, or increasingly often, because his sons 'don't want farming'. They've seen the stresses caused by the politicisation of farming and the endlessly encroaching stranglehold of the supermarkets.*

* Currently, supermarkets are rejecting up to a third of vegetables because they have decided they are not straight enough, small enough, large enough, pink enough, etc., and the farmer is forced into stressing his land ever harder with aggro-chemicals to try to make up for this huge and pointless deficit and to survive for another year. Thousands of people leave the land every year. I wonder where our food will come from in 2020? Mars?

Or sometimes it's just a retirement. Fifty years of unrelenting care and labour, in all weathers, every day of the year, take their toll on even the toughest and most determined. The back begins to stoop, the legs and hips stiffen up, and the farmer needs a stick. The heart is still willing, but the body's had enough.

So Mr Jones decides to sell. The good news is that there never seems to be a shortage of people who want to buy his land, piecemeal or en bloc, and the house will always sell, with any luck at an inflated price, to Someone from the City with too much money and no wish to get his hands dirty. So Mr Jones probably won't starve. But it's a sad day for him. He has intimate connections with his land and animals. He knows their every inch and foible. He is an expert, about to lose his fields of expertise.

He calls in the local auctioneers who delve into his sheds and drag out all the old baling machines and hayrakes; and rollers and turners; a trailer or two; perhaps an old pony trap; and a pre-war thraddle-twanger, whatever that is. All manner of clapped-out and superseded equipment and tools suddenly appear out of lean-to's and the back of barns. Each item is assessed for saleability and is either itemised as a Lot, or chucked into a composite heap: 'Pulleys Assorted'; 'Items Miscellaneous'.

Any half-decent equipment, like a superannuated plough, for instance, might qualify for a quick spray of paint (plus eyeliner if appropriate) and is hauled out to a place on 'the field'. Smaller lots, tools, chainsaws, etc., are sorted, batched, numbered and locked into a suitable shed. Household goods are similarly dealt with. Stuff which is suitable only for scrap (i.e. not even worth the attention of an impoverished smallholder) is hurled carefully aside for mass disposal. Livestock is dealt with separately, perhaps a day or two earlier.

On the day of the sale, friends and relatives from miles around turn up early and are treated to tea and sandwiches. Other options may appear, depending. The most attractive items in the sale will have been published in the local press and on big posters around the district for weeks in advance. The whole area knows about it, and people arrive to view the goods throughout the morning.

Then at 2pm or whatever, the auctioneer climbs onto the back of the trailer, heavily refreshed with tea or whatever, and the tractor hauls him into 'the field'. A quick 'one two one two' into his bullhorn and the auction is away.

'Lot number 1, gentlemen ... a Fordson 5000 tractor; 1600 hours on the clock. Tidy body. Excellent machine. Where shall we start? ... How about you, Mr Williams? You could use a decent tractor couldn't you?' The banter is gentle and pulls people in. 'I see Mr James has been eyeing this one up ... wedding anniversary present for Mrs James, is it?'

Two minutes later the Fordson is sold, 'for a fair price on the day', and we all move on to the next big item, often a smaller tractor.

The 'big tack' is all dealt with within an hour and then the trailer and followers move on to the smaller stuff. This is the bit of the sale that is of major interest to us. Things like scythes, sledgehammers, Victorian swede slicers and maize mangles, made from virtually rot-proof and highly decorated cast-iron: anything small and hand-powered, that might make life a little easier. That amazing wheel-hoe was a major find.

Now ones cost £50 or more and are not actually very well-designed. Ours was a belter. No one was ever as pleased with a new Ferrari as I was with that fiver's worth of push-along.

At the same sale we bought a very battered, low, wide, bench sort of thing for 50p, which I thought would go well in the yard as a seat or footstool. A neighbour told us one day that it was actually a pig-slaughtering block. Judging by the damage to the wood, it must have been owned by someone with a passion for historic battleaxes and a deep hatred of pigs.

After the small lots the posse moves on to the oddments and assorted. The auctioneer's level of banter moves up a notch or two:

'Lot 153: an antique hay knife* … two actually ... and a couple of hedge slashers ... a fiver? Two pounds, then? Anyone going to Cardiff?

'Lot 154: a quantity of tools, it says here. Quite frankly I've no idea what some of them are, and I've seen a few. Mr Davies … do you think that one with the row of teeth down one side could have been for skinning dinosaurs?

'Lot 155: twelve bicycles, past their best maybe, but attractively displayed in a loose pyramid, I think you'll agree. £1? Thank you ... to the gentleman in the splotchy blue coat and the busby …'

There's plenty of bargains to be had at farm sales. And everybody's happy because Lot 148: a broken pitchfork and two pickaxe blades would cost more to advertise than they are worth. They're also a great source of obscure spares for outdated but restorable equipment; again, items too cheap to

* Anything over twenty years old is automatically an antique in a farm sale.

be worth advertising. Everyone's a winner. Antique dealers revel in the occasional stained glass window and passion-raddled ottoman, and a local man snaps up every old handtool whose woodwork can be oiled and metalwork wiped over with matt black. He sells them en masse to traders in various cities. People buy derelict tractors to lovingly restore. (Every show has a parade of classic tractors.)

You queue at the desk and pay the man. You then have twenty-four hours to shift your purchases. Not a problem with a couple of rakes and an old scythe cradle that you plan to fumigate and hang on the sitting-room wall, but which inevitably gets left in a shed until the woodworm polishes it off completely.

By the end of the day the farmer's life has changed radically. The herd of Friesians he's carefully bred up over decades has gone. The farm is silent.

His lovely tractor, maid of all work, has gone, and all his ancillary gear, old and new. The farm is empty.

The sheds are cleared; the house is cleared. The farm is dead.

No doubt Mr Jones has a way of coping with this in a quiet moment later in the evening, but just now, what he has is the knowledge that all his friends have turned up and 'got their names on the book', meaning that they have made a point of buying something from sentiment, not just for a practical use. That will be his last memory of a lifetime on his land.

Time moves on.

* * *

The two biggest bits of tackle we bought at sales were a pulley that ran off the PTO (power take-off) drive at the back of the Fergie, and a circular-saw bench. The plan was to connect the pulley to the bench via a whopping great drive belt and thus be able to plank our own trees when we had hauled them out of the wooded cwm. It never happened.*

A helping hand

We felt we could use a bit of help around the place, seeing as I was so unreliable now, and out of the blue, help arrived.

Our first contact with John S was as a WWOOFer (see Chapter 26). He came for a weekend, all the way from Redhill in Surrey, and we got on fine. He seemed to enjoy the work and brought with him a deep knowledge of birds and moths and butterflies and things that crawl.

His plan was to buy a little smallholding of his own, where he could grow and sell organic veg, and wondered if he might come to some arrangement with us. Yes. Why not?

And so, a few weeks later, he turned up again with an old Austin Maxi stuffed with his worldly goods, and set himself up in one of our two caravans. The idea was that he'd stay for a few weeks, working three days a week for us in return for lunch and dinner and his week's lodgings, and on the other days of the week he would fend for himself and hunt for his would-be holding.

With any luck he'd find one within a month or two. No

* What was I thinking of? How was I ever going to lift *a tree trunk* up to the level of the saw? And how could I hold it steady, without the blade jagging on it and exploding into a confetti of razor-sharp shrapnel? Heavens …

such luck. One month stretched to two ... three ... four ...

In the end, John was with us for sixteen months before he finally found something that was more or less suitable. He'd had a couple of false alarms on the way, of course, being kept waiting and waiting by a vendor, and then being ripped off. He didn't have a lot of silly money, so this hurt.

Meanwhile, he worked hard for us and taught us lots of things about wildlife that we'd never dreamed of, and even became a local warden in a reserve.

He was a very easy man to get on with, but how he put up with me, I really don't know, and have never dared ask. Where John was phlegmatic and sanguine, and coped with all his setbacks with a bit of a snort and a laugh, I was at the other end of the spectrum. Ordinary setbacks, like machines breaking down *again*, would bring out the melancholic in me. And setbacks caused by the stupidity or obstructiveness of others would send me spiralling off into hissy fits of choleric outrage. Anne was used to it and took no notice (sanguine, you see; possibly phlegmatic). I never took it out on anyone else, but I can't have been that much fun to be with from time to time.*

So now we had a little help, just when we needed it most. Luck?

* Much better now, thank you, but BT can be relied upon to bring out the worst in me, every single time I need to deal with them.

– 15 –
Work Parties

Smallholders are sociable people and good at seeking out like-minded souls to exchange a few joys and moans with. Soon after we arrived here I was delighted to learn that this networking instinct had been locally extended into a semi-formal format called a Work Party.

The principle is that every second Sunday all ten or so members arrive at one family's spread and spend the morning tackling a job that is too hard, heavy, or just too plain boring for one household to take on. We muck-out cowsheds, clear ditches, fell trees, weed miles of veg and unthistle acres of pasture. Then we lounge about eating our packed lunch and exchanging tips and gossip before putting in another couple of hours then rushing home for milking or tunnel-watering or collecting the kids from Sunday school or community service.

I warmed immediately to the first Party we went on when I heard four husky check-shirted types (all blokes) jokily debating the relative ecological merits of burial and cremation. Opinion was split until a fifth party suggested that as meat was a good source of nitrogen, maybe the *ideal* solution would be to rotavate the dear departed in instead. The discussion then became more technical as they tried to work out the rate per acre. Practical folks, I thought.

We really enjoyed our Party Sundays, even when we spent five backbreaking hours humping and barrowing tons of concrete for the foundations of Howard and Jean's new self-build bungalow. But we had to leave the group, as we were travelling up to forty miles each way to the other farms (which all lay relatively close together), and couldn't afford the petrol. But we really liked the Work Party idea.

Within a couple of weeks we formed a more local Party from people we knew. It was nothing like the previous one, whose members almost all kept goats, for reasons that defeated me, and carried on long conversations about pedigrees and 'supernumerary nipples' and other worryingly satanic and capricious things.

Our Party was different, and became the most unlikely group of people I've ever been associated with. One member was a professional theatre scenery painter, who could make anything out of anything, and several things out of nothing at all (he drew the cartoons for this book, too), and his wife; one was an artistic lady and her husband who had made a beautiful garden out of a patch of scrubland; one was a kilt-wearing prep school outcast who was convinced he was the bastard son of an Earl, along with his blind, and very feisty wife; and one was, would you believe, a *Guardian*-reading dentist, and his wife.

See what I mean about unlikely? A *Guardian*-reading dentist?

We worked together remarkably well, considering that we were such a diverse group. Obviously A grated with B from time to time and C got on everyone's nerves most of the time, but we did all pitch in and got a lot of weeding and ditching done. Not everyone was suited to every job, obviously. The dentist was excused any work that might risk his fingers and wrists. The blind lady was excused roofing work and chimney cleaning. Not strictly true, actually; she once had to be restrained from carrying a large sheet of galvanised tin up an unsecured ladder on a windy day. We persuaded her that if she were to slip, God forbid, then we'd have to hang around for *ages* for an ambulance, and the entire morning would be *wasted* in form-filling and blood-transfusing and so forth. So she surrendered her independence to the general good. Good.

Interestingly, the crunch point involved Time Put In. Some of us had much more freeform and impetuous ideas of a full morning's work than others. Conversely, some of us were much more obsessed with getting our pound of flesh than others. Why should household *x* put in six hours when household *y* put in only three?

The resolution was masterly. We cut up chunks of vinyl tiles into tokens; one chunk = one hour. Everybody had twenty hours allocated, and off we went. Fine. For several months the system worked well. The Romantics tried a little harder, and the Industrialists could be generous, now that the principle had been established.

The trouble came with a pair of newcomers. Very nice people. He was an ex-copper from the Met; she a nurse. They were officially issued with twenty newly minted tokens each. The local alternative economy was taking a step forward. One year on, we'd be a power in Threadneedle Street; a decade from now, Presidents would come crawling...

But we had not allowed for the People Factor. What could possibly go wrong? It never occurred to any of us that anything could.

Any ideas?

Mass flooding of the token supply with Taiwanese forgeries? Refusal to pay up? Fisticuffs in the shrubbery? A hostile takeover by a rival Work Party?

Nope. What actually happened was that the newcomers came to one Work Party, worked hard, and went away. Fine. Then they came to another; ditto. And another. And another. Fine.

But after four months, they had accumulated unto themselves every single token we possessed … and never requested a Party on their own holding. We offered; we cajoled; we almost threatened; but there was always a reason why this week (or the next) was no good. In the end, our entire economy just … disappeared, into some kind of vinyl limbo, and I was from that moment convinced that economics was bunk.

But I still love the idea of Work Parties.

* * *

And I still don't understand economics, except that its 'Laws' are ultimately based on either greed or fear or both. I'm fairly confident that I have some sort of grasp of at least the basics of all the other Big Isms of the world, from religions to science, and from philosophy to mythology and the Great Themes of History. But economics? Not a clue.

The basic problem is that I don't understand what *money* is, and when you've said that, you're in trouble. The best I can come up with is: 'Money is an abstract concept (represented by cowrie shells or elephants' ears or odd bits of paper) to facilitate trading, and is worth *only* what you and I agree to it being *at the moment*.'

All very well in practice, but it really doesn't work out in theory, does it? Far too vague. Money simply has no intrinsic value at all. It's all a huge confidence trick that we all agree to take part in ... and somehow we bumble along. Too messy for my liking.

And I still don't know why our own little coinage system fell apart. The copper and his wife cornered all the coins, yes, but then what? Should we just have coined some more? But that would be inflationary, wouldn't it? And anyway, what if the copper just collected all the new issue as well?

I'm baffled.

* * *

Baffled or not, wouldn't it be nice if the Work Party idea spread far and wide? Imagine, every second Sunday, all over the country, from the wildest of the wilds, to the densest urban jungle – little groups of people coming together to help one of their members out with a big job. Spring cleaning whizzed through, rooms decorated, gardens trimmed, dug and planted, patios concreted, fences mended, walls painted, horses shot and stuffed. Cheap, sociable, fun, bit of a picnic, helpful – it seems to me to be a brilliant idea.

And, of course, skills would be gradually traded en route. Everyone would become a little less dependent on specialists and 'experts' who are inevitably too busy to be bothered with you when you need them, and too expensive to boot. In this age of the Internet and email, it would be a breeze to organise a local group. It just needs one person somewhere to think, 'Oo ... *I* could do that ...'

I wonder who that person might be?

Chainsaws and Tussauds

Part of the self-sufficiency gig, for us at least, involves growing our own fuel. We had plans for coppicing as I've already mentioned, but before you can coppice, you have to fell the big mother-tree from which the coppices will sprout. And to fell a tree you need a chainsaw. Even I knew that, after spending hours in Nottingham trying to cut up a twelve-inch-thick log with a rather dull bow saw. 'Appropriate technology is the watchword,' I thought (among other things), as the saw finally refused to move either forwards or backwards, or indeed, upwards. My only option was to unbolt the blade and leave it in the wood for someone to gash their legs on later. Serve them right.

But I knew nothing about chainsaws except that they were extremely dangerous and made a noise like a trials bike stuck in quicksand.

The only other thing I knew was that for once I wasn't going to buy second-hand. For such a dangerous item a Zero Risk policy was vital.

I researched as many brochures as I could find and asked all our friends and neighbours, who all thought I was being just a shade over-anxious.

In the end, I settled on a bright yellow Partner, made in Sweden. 'Swedish means reliable,' I thought. 'Look at Volvo; look at Abba. Also, Swedes know a thing or two about sawing trees. And – the clincher – yellow will be hard to lose.'

* * *

People are still killed felling trees. It looks easy, doesn't it? You decide which way you want the tree to fall, then cut a notch to force it to fall where you want it to. Then you saw from the opposite side to the notch ... and over she goes. What could be simpler? But what if the top growth is heavier on one side than the other (as it always is)? That will pull the tree to one side as she falls. What if it bounces off something else on the way down: another tree or, heaven forefend, your own house or car, or a dear little kitty, because you've miscalculated its height? Where will it end up? What if the top growth is wet after rain? Will one side be wetter and thus heavier than the other? That will affect its fall too. And what if it's just a tiny bit windy?

What if your cut isn't as straight as it could be? What if the grain of the wood is complicated by a burr or an outgrowth of some sort? Or if it's denser on one side than the other?

And what if you slip? A lot of trees grow in 'iffy' conditions: on rocky slopes; clinging to skiddy mouldy banks; near holes and small ravines in the land. And what if the cow comes for a cuddle just as you're stepping away from the toppling trunk?

Once the tree is felled, the dangers aren't over. Imagine: a fifty-foot tree, felled onto its side. You need to lop it before you can get at the trunk. The uppermost branches will be easy. The saw will buzz through them with little danger apart from the cloud of sawdust and the possibility of being knocked off your perch by a tumbling branch.

But the limbs on the underside are a quite different matter. They are supporting the full weight of the trunk and any branches above. They are under enormous tension. If you saw through them willy-nilly you are in for a very rude surprise at best. A three-inch branch whacking you across the forehead will give you pause for thought, possibly scattered in dollops across the woodland. An experienced local tree surgeon caught a full wallop like this once, as the tension on the branch was released by his saw. It knocked him stupid, naturally, but it didn't stop there. The force of the blow hurled him across the width of the A40. Thanks to the temporary traffic lights, he escaped alive, but with a very sore head and a revised Safety Schedule.

Anything bigger than three inches, released from the enormous tension and pressure, could easily kill you.

* * *

The only inherent danger in the design of the machine is 'kick-back', caused by forcing the nose of the saw too roughly into a cut. The saw might, well ... kick back, and upwards, in an unpredictable direction. All saws are fitted with an auto cut-off device, so the real risk is slight. I've never experienced it myself.

You can get all sorts of fancy safety gear for chainsaws, but I'm not convinced of their value. Ear-defenders? Mmm ... maybe, if you're working for hours on end, but what if someone shouts, 'Look out!' or 'Grizzly!'? Or worst of all, 'Tea up!' and you just don't hear them?

Helmets? Can't see the point. They'll be bound to fall off unless buckled on. And what would they defend against? A saw going straight into your head? How on earth did you allow that to happen?

There are fancy costumes available which are stuffed with fibrous material that will snag and clog any saw chain that wanders into it. OK for people given to narcolepsy or uncontrollable giggling, I guess, but should they really be logging at all?

* * *

At the end of the day a chainsaw is just a tool, like a drill or a sewing-machine. With common sense, and a bit of regular servicing, it's really not a danger in itself. Any problem is almost always self-inflicted – usually by someone miscalculating his stance and its relationship to the Law of Gravity. 'Awareness' is your best defence.

Just as a matter of interest, I do wonder whether 'safety' features in general are not just aids to complacency. Wouldn't road deaths come down to zero overnight if we replaced multiple airbags with a six-inch steel spike in the centre of every steering wheel? The 'emergency stop' during the driving test would bring home its purpose to all but the terminally stupid.

* * *

Actually buying a chainsaw felt like some sort of rite of passage, because only proper grownups could use one.

I read all the instructions ten times then put my foot across the guard and yanked on the cord. Twice. Then read the instructions again. Then switched the choke *on*, (whoops). Another tug and there she blows! Blue smoke and that stuttering putter-put ... put that can only be a chainsaw. Squeeze the trigger and ... whee!! Round the chain rips.

I am now restrained from absolute mayhem and a place in history and Madame Tussauds by only my repressed upbringing and perfectly hinged mind. Best hang onto that thought …

The ash log, some ten inches across, is awaiting. I hover the blade over the top surface, squeeze, and lower the chain of death onto it. Zhhhwiyizzzz … and the chain cuts through the entire log like a knife through hot butter. Ten minutes' sawing in ten seconds. This is *fun* …

Anne has always regarded our Partner with a sense of deep unease, but I love it. In that first year I cut a huge mound of firewood with it and gradually learned its ways. Beware of branches under tension; look out for kick-back situations; don't allow the blade to get jammed in the wood; go slowly; think twice, cut once; better make that 'think *three* times'; never work anywhere near animals; be very aware of your stance; be even more aware of your entire surroundings; never let the blade stray into a line with your leg; never for one moment let your concentration wander; and don't even *try* sawing across a big fork, where a major branch leaves the trunk: you'll be there all day, and it probably won't cut through it anyway. Surprising, but true. I also decided quite early on that I was probably never going to get good enough at it to take the saw up a tree. But it looks so easy, doesn't it … Perhaps just once, just to see …

* * *

The only time I've been genuinely frightened by a chainsaw was during a Work Party, when the son of the Earl did a bit of overhead lopping with his own saw. The chain was so loose you could stick your finger between it and the bar (should you choose to). One small jam or twist, and the chain was going to come off the bar completely and flail about, freeform: Extreme Ribbon Dancing. I kept well away.

– 16 –

New Leaf Project

Four issues:

- Everyone wants more and cheaper organic veg.
- Lots of people want to get out of the cities.
- Nobody is happy paying East Anglian barley barons millions to *not* grow crops.
- And everyone deplores the depletion of wildlife and habitat, and the depopulation of the countryside.

Could we combine these four problems, I wonder? And find a joined-up solution, perhaps thus:

Why not set up a Trust whose brief is 'to buy a 500-acre ex-barley farm in East Anglia, and to split it up into smaller units, whose purpose is to co-operatively produce lots of affordable organic food'?

The split-up might be into, say, 1 x 150-acre farm; 2 x 50 acres; 10 x 10 acres; plus 50 acres split into 20 lots of varying sizes.

Two things become apparent here:

1. We've lost a hundred acres somewhere (but read on), and
2. We're talking about not just a gaffer, a day-man, and several huge machines, but a population of at least fifty people. Who are these people, and where do they come

> from? Some would come from the ranks of organic organisations who have always wanted to try living off the land, but could never afford to buy in. Others would come from training schemes set up by various bodies, including government initiatives. And others would come from elsewhere, including Holland, if I know anything.

We're talking about building a hamlet, really. That's a very big deal, and expensive. Obviously the settlers would keep costs down by self-building, using appropriate technology and materials to minimise running costs.

Once up and running (and we don't need all thirty-three units to start on the same day) they produce organic food, as economically as possible. This means appropriate co-operation. Big Farm does the ploughing for everyone, for example, and small farmers help pick Big Farm's two acres of runner beans. Details TBA.

Marketing needs to be carefully thought through in advance by the Trustees and at least some of the farmers. Top priority would be to sell locally, including 'at the gate', then further afield, primarily to small shops. Some cautious arrangements might be made with some supermarkets for bulk crops. Value-added processing would be high on the agenda, particularly as a means of generating income off-season. Polytunnels producing salad crops might need to import labour, thus creating local jobs.

So let's assume all thirty-three units have negotiated their relationships, via a suitably mandated council of their own members, and are happily marketing hundreds of tons of lovely veg. In the evenings, if they so wish, they entertain and bore themselves and each other with charades* and Bruce Willis videos, or pitch into communal projects like laying new water lines, or working on the 'Big Green Brother' Video Diary to sell to the BBC as 'diversification'.

And there's more! The whole place should be a hive of experiment, training, and education. Sooner or later facilities would be built on the hundred acres that went missing from the calculation above, to house weekend learners from the cities; students on bursaries; paying guests; WWOOFers (see Chapter 26); visitors from university agronomy departments, checking on their projects; holiday-makers who are sick of being ripped off and bored rigid at the seaside; scout and guide groups; Portuguese Woodcraft Folk; gardening clubs on bargain breaks; all manner of people, all wanting to learn, and happy to sing for their home-grown supper (unless physically restrained, in certain cases).

More formal research and education takes place in the experimental plots, laboratory, and lecture hall.**

This is only the broadest of pictures, but you get the drift.

Would it work? Well, I'll guarantee that ninety-nine out of a hundred readers will shake their heads and smile at such naivety: 'It'll never work.' They may be right.

---

* The last time I got suckered into playing charades, you got your subject by lucky dip from a list of TV programmes. While other people got easy stuff like *The Bill* (project fingers forward in front of mouth; raise and lower thumb beneath them) and *Top Gear* (waggle a gearstick about in a non-suggestive manner), I got *Agatha Christie's Hercule Poirot.* Go on ... I defy you.

** If you think there is nothing left to research in farming/gardening, read Tompkins and Bird's two astonishing books: *The Secret Life of Plants* and *Secrets of the Soil.*

But I also guarantee that there will also be that hundredth person, the one with a dash of spirit, who'll say: 'Now *that's* what I call a worthwhile challenge.'

About fifty thousand people will read this article. One per cent equals five hundred live-wires. Not a bad start. Australia probably started with less.

So I don't doubt that thirty-three good wo/men and true will be findable. What about the money? Trickier. I've no idea how to cost this adventure, but I'm absolutely certain somebody does. That somebody should be on the Trust. Who else should be there? A worthy and competent patron with 'gravitas power'; also, the persons who prime the financial pumps, who might well be pop stars or lottery winners, or successful entrepreneurs, who are looking for a creative and worthwhile cause; and an organic guru. Other experts will be co-opted as necessary.

A foolish fancy that would just soak away millions? I don't see why, because the New Leaf Project would have a sound business plan, using its mountains of prime veg to pay its way and refund its set-up costs, if appropriate. What's more, it would gradually build up enough capital to eventually buy the 500-acre farm next door, and, guided by their own experience, would help set up a second co-operatively independent Project. Some years on, both Projects would buy a third farm. The scheme would grow … organically. Eventually the UK would become organically self-sufficient. And those 'four issues' would be resolved en route.

I can probably think of more snags and problems than you can, but I also know that the longest journey begins with the first step. And we all know this journey must be undertaken sooner rather than later. Why not now? Any views?

* * *

After this article first appeared in the HDRA magazine, about a dozen people contacted me with an interest in the New Leaf Project.

Two snags quickly became apparent. First, most people assumed that just because I'd mooted the idea, I was then going to organise the whole thing myself. A couple of bold souls did pitch in with some ideas of their own, but as we lived hundreds of miles apart, we couldn't really make any progress, even by email.

The second snag was much more important. I still think all the points I make above are valid, but they all hang on one thing which I didn't broach in any depth: ownership. OK, a Trust ... but what then? Would individuals rent their properties from the Trust? Would private ownership be possible in any way? If not, then people moving in might either lose their place on the property ladder when they felt they wanted to pull out for whatever reason, if they had sold up when joining NLP. Or, conversely, maybe some people would use the scheme as a means of finding a subsidised billet while they rented out their own house at a big profit. Problems.

If private ownership *was* to be allowed, how would it be administered? Who would be 'worthy' of joining? And could the Trust reasonably impose restrictions on who the new owners might or might not eventually sell to? Might some people just use the scheme as a get-rich-quick investment opportunity? And would the idealistic venture just end up as yet another island of middle-class consumerism?

No doubt wiser souls than I can suggest workable solutions to this problem. I'm sure there must be precedents.

Bearing all this in mind, think back: after reading the

article, did you find yourself thinking something like 'Wow ... that sounds fun!' before you thought 'It'll never work'?

I think an awful lot of people will find the basic idea appealing: a meaningful, human-scale, productive, and co-operative life. And once an idea is 'out there', if it really *is* a good one, somebody sometime will take it up and run with it. Whether it actually succeeds or not will depend on a thousand things, but most of all it will depend on a few individuals having the courage to think it through and get it moving.

There will be other snags, of course. *Precisely* who will be on the Trust? And why? What will the Constitution of the Trust be? Who will devise it? Who will select the thirty-three incomers? By which criteria? How will disputes be solved? How will the NLP 'vision' or 'programme' be developed? By whom?

I expect you can think of other snags as well. But none of them is beyond the wit of wo/man to solve.

Personally, I think you would need a strong democratic base to elect a strong manager, who would be the gaffer until pitched out on his ear by said strong democratic base. The danger I can foresee is of the vision becoming enfeebled by political manipulators and power-game queens. Some people are like that, you know. I would put a lot of effort into keeping the power-freaks away from power.

Another constant worry for the Trust, before the NLP could be set up, would be the problem of the supermarkets. I'm sorry to keep banging on about this, but they will do everything in their power to reduce the producers of veg to serfdom. They will promise all, then pull the plug, and leave you in a mess. Then, unless you have laid your plans well, you'll have to crawl back to them, on their terms, which will be beyond extortionate. Thus, the Trust would need to be supremely careful in its marketing strategies, right from well before planting the first seed.

Sensible caveats aside, many other people have set up successful co-operative ventures. A group of three villages in North Wales bought a mountain, to improve local job prospects. Several schemes are up and running in Scotland too. It *can* be done.

Pack-house and pool

After fifteen months of negotiation we finally got a Welsh Development Agency Rural Improvement Grant to convert a near-derelict barn into a much-needed weatherproof pack-house. The barn had begun life as a home-built milking parlour, tacked on to the house, but had since been knocked about something awful. The little window was rotten, the roof beams were rottener, and the 'roof' (a sagging mess of cracked and overlapping asbestos sheeting) was held on by inertia, and a few wisps of wire. Then one day someone thought it would make a good tractor shed and had simply hacked a huge hole in the wall. The Roof of Damocles wobbled over it, unsupported ...

Worst of all, the gable end had been exposed to the elements for so long that water had soaked right into the fabric of the infill, and frost had forced the inner and outer layers apart. The outer layer leaned out over the pathway at an alarming angle. You passed by that wall at your immediate peril (four times a day, rain, shine, or gale, on your way to milking and back).

Or so it seemed. In fact it was as safe as an angel's promise. We needed to tidy up the worst of the outward-leaning wall for the grant work, so we carefully erected our D-I-Y scaffolding and gave the top stone a tentative prod. Nothing moved. A good old whack, then? Still nothing. We eventually had to hammer and chisel all the 'dangerous' material off, one flat rock at a time.

The Plan was that we would do the place up to a standard that would eventually allow us to knock through into the house. The WDA was ahead of us … they knew that any shed with a huge picture window* was certainly destined to become an extra bedroom one day, and they made suitable specifications for the build. Thus, they insisted on roof beams that would one day take the weight of slates instead of the plastic-covered steel sheeting we currently had in mind. Slates? Those beams would take paving slabs or moderate sarsens. Huge things. They also insisted on a whopping great concrete ring beam round the top of the walls that would withstand a plane crash.

James the Builder and Chris, his labourer, turned up in the springtime and spent three months doing all the clever and heavy stuff while I did as much as I could of the mindless bits: mainly hacking out rotten lime mortar and slapping in endless tin-traysful of four-to-one cement. We're not talking regular brickwork here. The wall is a sort of jigsaw of slabs of schist, bits of boulder, and quoins of quartz. Not a level or straight line in sight. The mortar needed forcing in three inches deep in places. Endless, endless work, and surprisingly

* £1 for the aluminium frame, at a farm sale; £90 for the glass, a year or two later.

difficult. The experts tell me you don't trowel the new mortar in, you fling it. Well, I did try but ended up with mounds of damp cement all over the floor and inside my wellies and precious little in the gaps. Never mind. It got done, and I was pleased to have done it as the ME stopped me doing anything more adventurous. One night I went out to preen over the day's progress. On the scaffolding board six inches above my head was a dirty cup. I didn't have the strength to reach it. Unbelievable.

The new pack-house was a very welcome boost. Now we could barrow trays of cabbage and buckets of courgettes into the dry and process them at leisure. We bought some extra scales and looked around for a really sturdy table.

As luck would have it, one of the pubs in Newcastle Emlyn was changing hands (again). Anne knew I'd always hankered after a pool table so she suggested I ask the manager if he wanted to sell his. Astonishingly, he did. Even more astonishingly, he let me have it, balls and sticks and scoreboard and all, for £50. I dashed out to the carpark and drove the Bismarck onto the pavement. The Town Crier was passing by. 'Could you stop the cops arresting me for twenty minutes, do you think?' 'Certainly, sir.'

An hour and a half later we had the table assembled in the pack-house ready for a few test games. Very good, even if the cloth was more like satin than baize. Next, an 8ft x 4ft slab of chipboard to go over the top … with a remnant of (very ugly) floor vinyl stapled over it, and there we had our perfect waterproof work table. After hours, like any self-respecting speakeasy, the top came off, the lights went on, the elderberry wine came out, the boombox warbled, and it was Showtime.

The pack-house stank of stale beer and fags for months as the spirit of the Three Compasses gradually soaked out of the baize and away …

* * *

A great step forward. But nothing's easy, is it? The first frosty night showed us that rain is not confined to the outdoors in Wales.

The problem was that condensation formed on the underside of the box-profile roof and dripped on to whatever work might be on the table. This didn't matter too much for cabbages waiting to be humped into the Bismarck next morning, but it meant that tools and papers were constantly getting dribbled all over ... a rusty saw? A soggy boombox?

We tried sticking up rolls of 2mm polystyrene, but the condensation melted the wallpaper paste and the plastic just peeled off the ceiling like old bandaging. What's more, mice got into the little tunnels between the steel ceiling and what remained of the sagging poly and nested there. There was no way we could get them out, short of breeding very long thin cats with gecko-style feet, and we just didn't have the time.

Some years later we got another builder to bang up some studs which he infilled with one-inch polystyrene slabs, the same as I'd done to insulate the sitting room ceiling. He topped it off with plasterboard.

This is a great improvement, but it needs tidying up, and maybe even plastering, because it still rains in small straight lines from the occasional little strips where the bare roofing metal is still exposed.

* * *

Apart from the mice we get occasional problems in the pack-house with rats and voles, and, for all I know, lemmings and wallabies. Produce gets nibbled, and of course 'nibbled' = 'write-off'. Every now and then we lose patience and put down traps, especially when the pests reach a point when they defy the precaution we took after reading some sound advice in a smallholder magazine: '... always store root crops in strong paper socks'.

– 17 –

Autumn/Winter or Thereabouts

Some people define the beginning of autumn as the time the swallows leave, or when Manchester United have got the League-Cup-Sponsored-by-**Gleamo!**-the-Brighter-Choice-in-Kitchen-Hygiene!! stitched up, or the unspoken anniversary of Cousin Harriet's Big Mistake, or whatever.

For us, autumn is on its way when we find ourselves looking anxiously over our shoulders at jobs that will soon need doing urgently, and with faint regret at all the jobs that didn't get done in the previous all too brief summer.

The lawn, for example: yes, it did get a couple of rough licks, whenever we could persuade the mower to start, but now it looks like a neglected corner of rough prairie. Never mind.

There are also bits of tree everywhere, awaiting butchering and stowage. It is amazing how Nature slowly overtakes you. Suddenly you realise that the drive is hugely over-arched, and that a row of twelve-foot-high ash trees is sprouting between cracks in the concrete of the bottom yard. They seem to have appeared overnight. I noticed the other day that there is an elder trunk four inches thick growing between the brake cable and the frame of a retired bike. It must have been growing for years, unnoticed.

So, we lop and chop bits of tree and shrub every now and then. They are left to weather, then the big bits are stacked for firewood, and the twigs are snapped up for kindling. Meanwhile, they lie round the place in great untidy heaps. We should have done something about them earlier. Never mind.

The animals are pretty sure when it's autumn. The sheep line up at the gate, bawling for alms. They've got plenty of grass, but they want a bit of a cuddle after last night's drenching. OK, a cuddle; now go eat the grass.

April takes to spending more time at the gate, too, all winsome smiles and fluttering lashes, trying to shiver pitifully. Then one sodden day we go to fetch her, rattling her halter. She paws the mud, ignores the halter, barges past you and stamps off to her winter quarters, head down, at a brisk waddle. As soon as she's in, she grabs a duvet-sized swatch of hay and sticks her munching head contentedly through the window, to get a bit of gawping in before bed-time. Bliss.

So, yes, it's officially autumn when the cow comes in.

The outstanding job of the moment is to keep a close eye on the weather for sudden stormy bursts, especially at night. Sudden squalls mean the ditch pipes block. Also, big leaves seem to make a beeline for a little stretch of guttering over the bathroom roof. If you don't get up there and heave them out from under the overhanging slates the water backs up and eventually goes to earth via the bathroom ceiling, then floor, then mop and bucket.

The other job is to start collecting buckets of ash from the stove. Normally this gets scattered along the 'paths' across the back grass-patch ('Lawn'? Ha!), but as winter approaches we need to prepare for the arrival of the Eastern Glacier. A glacier? Surely not! Yes, I'm afraid so …

There are several little springs scattered along the drive

and yard. Normally they gently weep and are just 'there'. I like to think of them as a water feature.

But when the frosts arrive, these springs seep and freeze, and seep and freeze, until the whole of the lower yard becomes a single solid sheet of ice, several yards wide and long. A serious nuisance, because our carport is at the bottom of this slope. Once the glacier has formed, it is impossible to get the car out. In fact it is so slippery that it's almost impossible to get to the car in the first place. We take it in turns to try, while the other awards points out of six for Originality of Twists and Stumbles, and Silliness of Noises Emitted. Points are subtracted for time-wasting fractures.

Needless to say, the Glacier is the last ice in the entire district to melt, as it sits in a pool of deep shade. The car, and hence we, were once marooned for three days because of it while everybody else was up and running after the thaw. I'm sure nobody believed our tales of crampons and crevasses.

Perhaps the smart thing would be to drag all the half-dried shrubbery onto the Glacier and burn it there. End of two problems. Possible end of car as well, of course, so perhaps not. We'll just stick with the buckets of ash, I think.

* * *

The weather in West Wales is pretty moderate because it sloshes over us from the Atlantic, rather than slicing in off the Siberian tundra. We don't have much in the way of snow and ice.

Light frosts are relatively common though. I love to see the crystal feathers on the gate, gleaming and scintillating; and the suede baize of powdered diamond on the car roof. Why the baize and not the crystals? And why did this particular puddle freeze across in this particular pattern of straight lines and planes? Why not another pattern? And if all the molecules are in random motion, as they say they are, why straight lines at all?

In the suburbs, we used to do as others did, and regarded frost merely as a damn nuisance that froze doors up and crazed the windscreen. Out here, it's a part of the great outdoors, and within reason, not a problem. The animals' water freezes over, so you keep an iron bar and a thick glove handy to smash the surface ice on the bath or bucket. The problem arises when it freezes really hard and the bucket freezes right through, and the spring supply pipe does too, unless you've thought ahead to leave the tap trickling overnight. The only answer is to ferry out buckets of warm water to those in need. Of course, they won't drink it because it's *warm*, and hence unusual, and hence unthinkable, but they do have the option. More than that we cannot do. After a few days the frost goes, and the only problem remaining is the Eastern Glacier, and an occasional rock-hard strip of ice that unaccountably forms on one side of the drive, just as it's beginning to level out. It's *just* enough to spin the differential and strand you.

Winter Field

dull jaded blades
spiked with stiff stalks
muzzed with moss.
one sheep,
separated, lost,
rushes back and forth
distracted
cold, wet, thin
and
howls

* * *

But once in a while we wake up to an inch or two of the fluffy stuff and the local kids make the best of it. Only the super-rich and the unimaginative have any use for a sledge. *Real* kids use plastic fertiliser sacks. Being organic, we didn't have any, but Edward from next door had a useful supply. He and Cait would trog off to the steepest slope they could find (one of Edward's top fields was favourite) and take turns to hurtle down it.

After a while the super-smooth plastic impresses a super-smooth channel into the snow, like a miniature Cresta Run, especially if you first half-pack the plastic sack with snow, to give you a perfect saddle shape. You scrunch the top of the sack together with both hands to stop the snow falling out, then lean right back to get a good sheerline at the front. You do a couple of preliminary runs to lay down a basic track; and then you are ready for the real fun … *of trying to barge your way through the hawthorn hedge at the bottom of the field.* We, as parents, knew nothing of this ambition, naturally. To us, the kids, aged nine or ten, were having a merry Victorian Print sort of a time, gaily slipping and skidding on the pristine snow with ruddy cheeks aglow, muffs and pom-poms everywhere, and innocent peals of laughter mingling with the merry bleat of cheery snow-dappled sheep. The true ambition, of battering a way through a hedge bristling with eye-ripping spikes and hidden whorls of rusting barbed wire remained a secret. All was well, so all was well. Mercifully.*

* * *

We thoroughly enjoyed the brief respite of a day or two of snow. The sheep coped with it fine, especially as they got double rations, and April coped too, by sticking her head out of her window and gawping as hard as possible at it.

It was amazingly silent. Perhaps a buzzard would keen, or a crow might hack, but otherwise the only sound would be of childish squeals of delight, echoing back and forth in the still air from miles around.

Under a quilt of snow, the fields take on a strange uniformity on both sides of the valley, dappled with tones and shadows cast by the low sun, the separating hedges drawn irregularly between them, like Chinese brushstrokes, each stumbling line suggesting its own story. Why does this hedge go from here to here? Why is that hedge out of alignment with the one opposite? What did that line suddenly turn left to avoid? Why is that field so little and triangular in shape? The

* Guess what they did in the summer, when there was no snow to tempt Fate with? More on this later.

entire social history of the district is laid out in black and white for those who have the eyes to read it. Every kink and quirk has its reason.

Even boring grown-ups have been known to take time to walk in the snow, just to experience the pristine otherness directly. Or maybe to see what can be seen ... spoor of blackbird, robin, crow: and wing marks on the surface where the bird has taken off. Deeper pugs of rabbit, fox, and polecat. Where do they come from? How many are there? Information to be borne in mind for when yet another half dozen chickens are killed by yet another fox assault.

* * *

The only real challenge the snow brought was the hair-raising stunt of driving the Reliant down the Mohican strip of geological crystal that builds up down the middle of half thawed roads.

If you're a fan of extreme sports, may I recommend driving a three-wheeler at 'traffic speed' along a major road after a decent snowfall has been mushed and crushed into a deep and crispy central ridge of very slippery ice? Forget free climbing; forget the luge ...

Hunting with cats

We had an episode with the local hunt about now. Half a dozen of their big friendly gallumphing dogs got lost (as they do every time; not terribly bright) and stumbled up from the cwm, making a row and frightening the animals. The sheep scuttled off in a bunch and April jumped the fence and landed in the vegetable patch. She meant no harm, poor dear. She was just panicked. Could have broken a leg. It happens.

We rang a few people and someone came round to collect the dogs. He wasn't very apologetic for the alarm they'd caused. It seemed it was actually my fault for not understanding The Countryside.

We're not impressed with the arguments put up in defence of hunting. For a start, if they were actually effective at killing foxes, they would have put themselves out of business centuries ago, wouldn't they? None of the other arguments hold water either, in our view. It all boils down to the fact that certain sorts of people enjoy killing things for fun. Let's just admit that.

But why use big daft noisy blundering dogs? What's wrong with cats? They are silent, single-minded, endlessly patient, and totally professional assassins. They are also unlikely to get lost, even though they don't have the astounding nasal abilities of hounds, like the ones who keep getting hopelessly bewildered in our valley, for example.

Any Master of Cats could breed up some proper whoppers, instil the basic lore of co-operative hunting into them (possibly a bit time-consuming, I must admit) and there you go: instead of twenty barking buffoons, you go out with five or six primordial predators, carefully woven into one integrated killing-machine. No fox would stand a chance. Inside a month they'd have killed every fox for miles around, and would earn every poultry-keeper's undying gratitude. (Farmers of grass or arable, of course, regard the fox as an ally in keeping down the predations of hordes of rabbits, so they might not be quite so keen on its eradication. You can't please everybody, eh?)

Speaking of cats, John the WWOOFer reliably informs us that the domestic cat is responsible for the death of millions of garden birds every year. This is surely an outrage? If my dog goes round savaging other creatures there is a hue and cry and hooded protestors paint 'murrdrer' on my gate. Fido is doomed.

But nobody seems to care about the wholesale slaughter of beautiful and pest-eating birds by pampered domestic cats who don't actually need the calories. Ikkle cuggly kitties? No … savage exterminators.

I suggest that cats should be treated the same as dogs, and be required to be kept on a lead at all times. Nobody will take this seriously, of course, as it 'goes against a cat's nature', to which I reply that I've yet to meet a dog who clearly prefers being on a lead to being let to run free as Nature intended, extruding offerings onto his neighbours' doormats and peeing on their alloys. Come to think of it, I've still not met a human who thinks it's fair that he's not allowed to park his SuperTurbo 6.6 17-valve 4x4 'Avenging Angel' exactly where *he* wants to, double-yellow line or not.

The price of civilisation is some degree of personal constraint. We accept this, on the whole. So do dogs, like it or not. I think cats should, too.

Some owners do accept their responsibilities, and I salute them. Their cats have a bell round their necks, or an electronic bleeper that gives any bird a half-chance to escape. But it's very hard to know how effective these devices are. The only bell that would be sure to be effective would be a triple-chimer, and weigh a couple of kilos.

To house-owners who are fed up with other people's mogs killing off their bird friends, might I recommend one of the new breed of high-powered water pistols, available from any toy shop? They pack quite a punch. Charged with a harmless but pungent solution known to be thoroughly offensive to cats (and their owners, I suggest), this precision machine should act as a powerful deterrent and give people their gardens back at last.

A householder might even put up a sign in his garden: 'Cat-free Zone. All Intruders will be Sloshed'.

In case you might be thinking baleful thoughts: no I don't hate cats. I've kept them and like them. It's what they *do* that disturbs me, especially in suburban gardens. It's a question of respect, surely? Respect for bird-lovers. And respect for the eco-system and the environment.

My Cat

My cat sits just so,
Essence of prim
Mouth like a teddybear
Toes together, just so
Tidy to a T.
Ears filtering the air
Featherboa tail, just so
Prettily over neat feet
Eyes re-charging in the sun
Dainty as a doily, just so
Completely self-contained.

Dreaming of murder.

– 18 –
To Have Bees or Not to Have Bees?

Picture the scene: a sweltering July afternoon; scented air; drowsy cow admiring the vista; bees buzzing; quite a *lot* of bees buzzing, actually; sheep nibbling … Bees buzzing quite alarmingly in fact …

'Dad! Your hornets are swarming!'

Action, not to mention panic, stations. Reading from the textbook in one hand, Dad is hauling on his venom-proof trousers with the other, desperately trying to remember everything he's book-learned over the last nine months. '*When the swarm eventually settles on a low branch, find a cardboard box and a table cloth. Direct a smart blow to the branch and catch the swarm cleanly in the box. Cover with the tablecloth and carry carefully to the prepared hive.*'

Yes, we have a box; yes it should be big enough; tablecloth? Thanks, Mum … What else? Prepared hive? Yes. Low branch? We'll see …

Dad is now securing his vampire-proof hat, with broad brim and fetching black veil. He can't wear his specs in there as they instantly steam up, so I have to read to him: '*How To Handle The Swarm: there may be up to 60,000 bees in a swarm* …' Sixty THOUSAND? … Each with a poisoned dagger …? is what impresses me most.

Dad meanwhile is forcing his hands down the black

rubber gloves which swell alarmingly with air at the fingertips, like a pantomime Aberdeen Angus.

OK. We're off. Dad waddles purposefully out of the house, clutching his cardboard reliquary and tablecloth shroud. I run on ahead and take up station in the tool shed to call out instructions from the book if necessary, and generally keep an eye on things: to ensure, for example, that he doesn't fall down the cwm and lie there undiscovered for a fortnight.

Ah yes, the cwm ... The hives are at the top edge of a patch of fallow land that slopes gently to a near-vertical drop of several feet, which in turn gives way to the stump-stubbled and bramble-knitted turmoil of the valley down to the stream.

'Fallow land' in this case means twenty yards of waist-high brambles, punctuated by swaying fronds of six-foot nettles, all held firm by bindweed. Beneath are generations of abandoned tin sheets, plastic tubs and tractor tyres. The SAS Training Officer would write it off as a death-trap.

And there, on the bottom edge of this wasteland, hanging onto a scruffy little elder tree, is the pulsing globular cluster of the swarm, humming like a cracked transformer: an alien being, if ever I saw one.

To reach it Dad must hack and heave his way through that twenty yards of Mother Nature's finest, carrying a box above his head, and sweltering in what amounts to a canvas straitjacket with matching trousers, in a temperature of 87 degrees. The good news is that the swarm *is* on a fairly low branch, and the 87 degrees is in Fahrenheit.

It's getting hotter by the minute in my shed, but nothing would induce me to open a window. I know those bees for what they are.

So I sit and watch the brambles snag and tentacle at his trousers. Every few yards he stops to heave them back up, with moderate success.

The collection is a snip. Whack and thud and that's it. The alien is snared. Astonishment. So easy? Then begins the long haul back. It's now broiling hot where I am, and I'm not super-wrapped and carrying a ball of death uphill, blinded by sweat, through nightmare undergrowth studded with snares and mantraps.

Slowly he advances, but has no hands free to save his trousers now. Buzz Aldrin meets Jacques Tati.

At last, hobbled at mid-calf, he reaches the hive. I call out from Chapter 9: '*Place a ramp at the entrance to the hive and pour the swarm gently onto it. They will march up it like a troop of little soldiers.*' Then: '*It will speed the process if you select the queen and put her in first.*' I don't bother reading that last bit out in the circumstances.

And … They do it! Amazing! three hundred and sixty thousand tiny little feet …

Dad is fascinated, hauls his pants up half a yard, loses concentration, and leans backwards … into the other hive … and knocks the top box off centre.

Suddenly the air is filled with another sixty thousand screeching bees, pouring out like smoke from a dragon's pyre. They are, quite frankly, vexed, and hammer straight into Dad at every available point. At times he is only just visible as a figure. His veil is blocked; he can't see; the noise is tremendous. Is the suit as proof as advertised? The gloves look very suspect.

He stumbles off and out of view, pursued by Hate. There

is nothing I can do. To go outside is literally suicidal. I hope the kids aren't in the field.

Later I discover that he somehow clambered and stumbled over two five-barred gates and ran, if that's the word, to the furthest boundary of our fields, swatting and swiping at every step. Finally he sank down exhausted and the bees dithered off, presumably to boast to each other over swigs of nectar and gobfuls of propolis how they done for that bloke in the hat good and proper.

That evening Dad drove us all the eighty miles to Swansea and back so Paddy could collect a maths prize. He had twenty-three stings on his hands and could barely hold the wheel. And, do you know, he kept those hives for several more years. Funny old world.

* * *

I couldn't get on with bees. 'Alien' really is the only word. Were they sixty thousand individual creatures, or just one? Whichever, they were savage/s.

* * *

Apart from bees and goats, another creature Anne and I never thought of stocking was a horse, partly because we had no spare land for an animal that couldn't realistically earn its keep, and partly because my two previous experiences of horses had left me permanently scarred (inside, at least) ...

As a sixth-former I tried showing off to my girlfriend by entering a local rodeo competition. Surely I could stay on a wild Welsh hill pony long enough to win the paltry prize? Well, no, I couldn't, actually.

I did everything right: climbed onto its bare back in the crush pen; gripped tight with my knees; grabbed two big handfuls of mane; whispered calming and confident words into its steaming ear; and presented my best profile to My Lady, lance couched and trimmed with her fluttering favour.

While I was perfecting my Sir Elvis leer, The Swine opened the box and the wild pony* blazed out into the arena like a rocket-fuelled Fury, trying the while to kick me on the forehead as quickly as possible.

I didn't have a chance. Before I could count to three I was sailing through the air like a human cannonball and crashed inelegantly to earth an hour or two later. The Swine batted the Fury off with his hat (no, not even a bravura display with cape and sword) and nudged me with his boot. No, nothing broken. But hang on – thumb at a funny angle. Dislocated.

'Am I going to win the prize?'

'Two seconds? Don't be silly, boy. That thumb's going to hurt, by the way.'

My second experience was the clincher. Anne's sister was a keen rider and once, when we were visiting, she convinced Anne she should go for a hack with her. 'What about you, Chas?'

Oh dear … a challenge. I protested that it was my clear understanding that every horse's ambition was to kick and bite and scratch you to death as soon as spit at you, and anyway, I needed both my thumbs.

'Oh, you can have old Tom …' (pronounced as '*Dear Old Tom*').

* 'Wild'? Positively demented.

Well … there's a dilemma, isn't it? On the one hand my thorough knowledge of equine malignancy, and on the other a bit of pride. And Old Tom sounded as though he might well be inclined to buy me a pint afterwards and tell me tales of the Boer War.

'Er ... OK, then. Thanks.'

The stables offered me a choice of two helmets. One was too small to be even forced on by a couple of whacks with the butt of a riding crop and a haul on the straps; the other was slightly too large, and thus rested more on my ears than my head, and rocked back and forth onto the bridge of my nose. As they tightened the straps, I felt my ears buckling. Not a flicker of a smile crossed anyone's face. In fact everyone looked once, and then turned away very quickly. I thought this might be some sort of omen.

The hack started nicely in Steve McQueen style, with a stately promenade along bosky Sussex lanes, but then swung suddenly left into rough and dingly woodland, with little tracks though dips and dells, and bracken up to the horse's throat. At the bottom of one of these dips, Old Tom stopped. Jackie was in front and Anne was behind.

'Give him a gee up.'

'Squeeze him with your heels.'

'Slap the bridle a bit.'

'Speak nicely to him.'

Obviously none of these worked. I might just as well have offered him a toasted cheese sandwich, because Old Tom was in charge for once, for the first time in years, possibly decades, and he knew it and was going to make the most of it. 'Oh come *on* Tom …' Jackie urged, but it was pointless. Impasse.

Then the two sisters formulated a plan. Accordingly, Jackie tracked back to a point behind me, then surged forward. As she passed, she leaned across and heaved on Old Tom's bridle, while Anne whacked him across the backside good and proper.

This woke him up and he hurtled off up the slope. Jackie let go of the bridle before I could grab it so I was being borne along uphill virtually bareback (shades of the rodeo came to mind almost immediately, you know), leaning back in the saddle like a novelty act in a struggling circus. Then, as the land levelled out, my feet lost contact with the pedals and I was bouncing about like a duck in a waterfall. I grabbed desperately at the bridle and tried to heave him to a halt. He was having none of it. Years of being 'Old Tom' melted away, and for a few moments he was Bucephalus and Copenhagen and Red Rum, all together in one glorious moment of thundering gallop. His only problem was the sack of bouncing turnips thumping into his backbone.

With the urge for vengeance born of years of servitude he did the obvious thing and veered slightly off the straight and narrow and headed for a low oak branch. It was only a foot thick, but it would have done for me what a windscreen does for a fly, bucket-sized helmet or not, especially as one foot had now somehow caught and lodged in one of the shoe-hoop-things and I was thus partly re-attached. I automatically did what Red Indians were always doing in cowboy films,* and leaned sideways across his neck. Death averted.

The cavalry arrived a few seconds later, as Old Tom, *Dear*

* I've been trying to remember the name of Roy Rogers' horse... 'Ginger', was it?

Old Tom, finally ran out of oats. Jackie got there first and hauled him to a complete and heaving halt. Harrumphs all round.

'You should have kept your feet in the stirrups.'

Nobody asked if I'd enjoyed it, for which I was grateful. In retrospect, however, I wish I'd had the grace to see it from Old Tom's point of view and taken my leave of him with a respectful bow and a shake of the hoof. But that other bough was still too close to my eyes for that.

* * *

I've had as little as possible to do with horses ever since. The last (rather surreal) time occurred many years later when Dad phoned down from his bungalow to tell me that seven horses assorted were wandering round his garden. Could I help? Yes, of course, I'm on my way.

What was I going to do? Yee-har! them out and away with a sweep of my gleaming white stetson? Or whisper them into a silent huddle in the potting shed? Actually my plan was:

1. Go up the hill to the bungalow.
2.

But before I could leap decisively into action, Dad phoned down again, 'They've gone.' (a pause) 'I suppose you think I've gone raving mad?' 'Not at all,' I replied. 'The fact that I will find no horses in your garden will be proof enough that they are no longer there.'

And true, there were no horses. But they had definitely been there. Hoof marks all over the lawn, and a near miss on the rhubarb patch.

I went home again and rang our neighbour to tell him his nags had done seven runners.

The East Wing

The year after getting the pack-house up we had an unexpected windfall of several thousand pounds from Anne's Dad. What to do with it?

I was keen to buy a neat little Kubota rotavator that would fit behind the Fergie and enable us to cultivate the field crop beds in a single pass. A hi-tec tool that would save a lot of valuable time and depleted energy during the spring panic.

Anne, however, thought we should use the money to build an extra room that her parents could use as a holiday base.

Obviously, this was the better option. It was only fair, after all. What's more, it would last longer and would be much more versatile: an *asset*, in fact.

James the Builder came round and we knocked up a plan. That is to say, Anne and James knocked one up. Actually, Anne told James what she had in mind and James made it work: if we knocked down the coalshed and shower room we could build a decent-sized bedsit, joined to our own sitting room by a corridor which could become a kitchenette. At the end of the corridor, we could incorporate that peculiar little bathroom that seemed to have been tacked on by a previous owner for no very good reason. A pair of french windows would give onto some steps into the drive.

A little bit of vision had turned 'an extra room' into a self-contained flat. Anne would have made an excellent architect.*

* * *

* Also Office Manager, vet, landscape designer, lawyer, Finance Manager, and Director of Education, though probably not all at once.

So one morning Chris the Shovel spat on his hands and picked up the sledgehammer with the sort of creative glint in his eye that we mere mortals can only dream of. Whack; wallop-wallop-wallop … wallop ... ther … **wummmp.** Down came the coal shed. So *quickly.* It wasn't going to be sorely missed as it was in entirely the wrong place: at the opposite side of the house from the stove, one door and two gates away. (*Why?*)

My major memory of it is of one damp autumn evening, after a hard day of wood-sawing and general hacking, and having to shovel a ton and a half of fuel from where it had just been dumped in the yard, into the shed; and all before it got dark or I could have my dinner.

'Only a ton and a half? Could be worse,' I hear you say. Yes, it was worse. The fuel wasn't the crisp and glittering anthracite we would have wished for but could never afford. Instead, it was the coalman's equivalent of turkey-neck offcuts: a weird material called pelemond, which seems to be some sort of 'washings': all the dust and clay that settles at the bottom of a tank. They hack it out with a JCB every decade or so, and sell it as 'solid fuel'. Well, 'solid' is fairly accurate, if 'fuel' is somewhat less so. Once dumped on the yard, it settles into a rough cone of virtually impenetrable *stuff,* halfway between extremely stiff cement, and granite. It is incredibly difficult to even get a shovel into it, never mind lever and snap a bit off and lift it. And being particulate, it is very liable to wash away in a half-decent mist, never mind a seasonal downpour. Hence the need to do it straight away ('damp autumn evening', remember).

As a fuel it is inadequate. The best it could manage for us

was a sulky glow and the heat it gave off was minimal. But it *was* fuel of a sort, and cheap, and we were grateful for it.

The shower room was more of a loss to us. It was actually very nicely laid out and had obviously been built for people holidaying in the caravans. But Chris was on a roll ... whap whap *wallop* wh … James and I held him back by the elbows while Anne rescued the window and the shower unit for future use.

The actual construction of what was to become the East Wing went very smoothly, although there were the usual alarums and excursions one expected when working with James. Whereas most builders seem to be of the phlegmatic persuasion, James was mercurial. Very competent at this trade, and thoughtful and imaginative, he was liable to sudden panics and depression when he thought something wasn't going right. Chris was a wonderful foil.

'Oh no! Oh ****. Oh ********** *********** h*ll.'

'What?'

'Oh **** ****.'

'What is it? James? *James!!*'

'Ohhhhh … it's all going to be out of line if I do this… ****.'

'Looks alright to me.'

'Yes but … see here … it's all … Oh … wait a minute … perhaps if I …'

And we were back on keel again. Never a dull moment working with James. More than once he had to sit down to recompose himself after nearly stepping backwards off the scaffolding to get a better view of his work. Chris, as ever, was a tower of support. 'Concentrate, James,' he'd say helpfully, in his baritone Manchester rasp. 'Here …' and he'd flick the radio back to Radio 3. His own preference was Radio 1, and they'd take turns, switching when one couldn't stand the other any longer.

And no, they weren't gay. Just a very effective working team. The main thing they had in common was enjoying their evening pints.* Otherwise, chalk and cheese.

'Chalk and cheese': quite a good description of the consistency of that pile of pelemond, come to think of it.

* *James:* 'Drunk? I'm not drunk. I can still *see*.'

– 19 –

Plugs, Plugs, Plugs

Disasters come in threes, they say. And presumably so do triumphs, although you tend not to hear so much about these. Perhaps it's because there are more disasters than triumphs in the world, or perhaps I'm just mixing with the wrong people.

Either way, this belief in triplitudes has all the makings of a self-fulfilling prophecy, if you think about it. Event A (say, a grand piano falling on you) occurs at Time A. You dismiss it, along with others of life's vagaries, like accidentally filling the kettle with milk. Don't laugh. But then, if Event B (say, an ACME safe falling on you) occurs soon afterwards, you associate it with Event A, and go round telling your friends about the odd coincidence. But then, because you know about triplitudes, you're on a constant lookout for the third: possibly an anvil descending from a clear sky. And sure enough, a mere fourteen months later, *almost* to the day, a leaf catches you a glancing blow on the shoulder while burying a duck in an autumn thicket, and Bingo! The mysterious triplitude has struck again!

But my own recent triplitude occurred over just five days and concerned, of all things, spark plugs.

Our rotavator ('rotavator', by the way, is the longest palindrome in the English dictionary. Not a lot of people know that…) is usually very good at starting after a winter's hibernation. Two tugs and you're away, and can bank on full power until fuel runs low, or, more realistically, out. But this year it puttered to a halt after struggling through a mere hour's work. I'm not a gifted mechanic, but I know enough to have an 'In Case of Breakdown' algorithm:

Step 1: Kick tyres.

↓

Step 2: Check petrol.
If OK so far, then

↓

Step 3: Despair-&-Panic.

Works every time.

But it looked like rain, and as a hundred yards is a long way to push a slumbering rotavator, I had to improvise, and remembered the popular ritual of Looking At The Plug. I'd tried it before on other machines and it had always been pointless, even when I could find the Mole grip, but I tried it anyway.

And lo! ... the plug was almost totally gunged with black treacly ... gunge. Part of it had burned to a crisp, forming a perfect warty carbon bridge between the electrodes. The plug was comprehensively short-circuited. Oh joy! A fault I could understand *and* do something about! It took a good ten minutes to clean it with a little wire brush and a set of dental picks I'd bought in the market months ago because I knew they would come in handy one day. And so they did. 'This won't hurt a bit,' I murmured, and hacked and gouged at the evil mess. 'Going anywhere unpleasant on your holidays this year? What are your political views? Personally, I'm a fascist. Great admirer of Doctor Mengele. Quick rinse.' The machine started first pull, and I was as full of myself as the dog when he retrieved a dead hedgehog.

Four days later, and at last a brilliant sunny day. Time to trim the lawn. The mower wouldn't start. No real surprise here, as the mower has always been as bolshie as the rotavator is willing. No point in kicking the tyres (solid), and the petrol was OK as I'd only just overfilled it, and had a wet shoe to prove it.

'Step 3: Despair-&-Panic' was gloomily descending when Anne pointed out, rather too chirpily I thought, that Argos had new mowers for £400. £400! Scandalous! Anyway, there's nothing wrong with this old machine. It just needs ... er ... Looking At The Plug! Yes, that's it. Piece of cake.

This time the plug was clean enough, but lo! ... you could easily trot a small *Clatteropterous* between the electrodes. *Far* too wide a gap. Even I could see that. So I whacked it with the Mole grip and put it back. Started first pull. I mowed away with that wonderful feeling of competence that drifts so rarely into my life, wondering the while how the gap had widened over the winter months. Metal fatigue? The tidal pull of the moon? The obvious answer was that it had been over-gapped all last year. But it had worked perfectly then, so why not now?

Never mind. Looking At The Plug had worked. And not just once, but twice! Oh, pride of great achievement!

Oh hubris!

The following morning, when the car went for its MoT,

Iori the Lorry was not impressed. 'Your plugs are loose, man! Rattling! Aluminium engine ... they'll blow out like mortar shells! Whoever put them in should be shot!'

I don't think he noticed me shuffle my feet.

* * *

Two postscripts: Alun the Mechanic pointed out that the reason the mower plug gap was so large was that the inner electrode had just burnt away a bit, like a wick in a candle. That's what plugs *do.* Ho hum.

Some weeks later, Alun changed the plugs on the car. 'Did they come out easy?' I asked, knowing that Iori had tightened them up like the bolts on a royal chastity belt.

'Easy? They were rattling, man. They could have shot out and killed someone.' Now what do you make of that?

* * *

Clichés hold a certain fascination, don't they? At school we are told to avoid them like the plague, but why? The only reason they are platitudes is that they reflect a truth so commonplace that everyone recognises it. A platitude is a sort of stereotype. And where would we be without stereotypes? We need our boxes and labels to make sense of the world, to enable us to cross-refer experience with memory, in drama and art, and in real life itself. And one of the odder platitudes of real life is the idea that things happen in threes. Why not fives? Or ones? Who knows?

But threes it is. The peculiar triplitude of the spark plugs was a bit odd. But last summer the Curse of the Threesome struck a second time. Curiously, it involved the same old

Mountfield lawnmower which had featured already in the Plugs affair.

Thus, within twelve days:

1. It's an old engine and a bit of blue smoke is no novelty. But grey-blue smoke *is*. So is orange flame. Dry grass in the exhaust was the problem. Much beating and kicking at it, and leaping about shouting and pointing, solved the problem.
2. An excursion under a rose bush knocked the spring off the carburettor. Strong stink of petrol. Luckily no dry grass or orange flame this time. But there *was* a loud cracking and splintering sound. This was caused by the blade mincing the cover to the air filter which also had been snagged off by the thorns.
3. The starting string snapped. Big tug; full-on effort … Snap. Stagger backwards at some speed, alarm in the eyes. Slip. Fall into comfrey patch, still holding the string handle. Oh, jolly amusing. But at least I hadn't been shrouded in flame or had my boot minced.

* * *

That engine ended its days when I noticed it was smoking a bit more than usual and realised I'd never put any oil in it since it had been passed on to us several years ago. Criminal, I know. How could I have done this? Too busy to spare the time, I suppose. And anyway, it was only used for about ten hours per season, so …

So … I drained the fuel out and tipped the mower over. Underneath was a dense mass of botanical fur and felted grass.

It took twenty minutes to hack it all off and carefully remove the lace-like skeleton of the exhaust pipe. Ah! There's the drain plug. Eight spanners later I drained out two full dessertspoons of molasses. I was chastened. How on earth had that machine kept running if that was its only lubricant? Shame on me.

With the happy heart of the competent nurse and mother I cleaned everything in sight and filled the sump with shiny fresh oil. I even removed the blade and sharpened it to cut-throat standard, for the first time in its life. *And* I fitted a new spring washer, and ran the oil can over every joint and bearing I could get at.

I heaved on the new starter string and away we went. Shaggy grass turned to rich velour before my eyes. We were a team! I had cared for this motor at last, and it was responding. My heart warmed ...

Ten minutes later it stopped dead, and a tug on the string, slack as a vicar's handshake, produced only a little 'clink'. Busted con rod.

THE END.

It took four months to find a replacement engine, as the Foot and Mouth restrictions of the day meant that the mechanic couldn't get to all the sales he wanted to. Meanwhile Dad lent me a couple of his own retired machines. One was a neat suburban cylinder job, excellent for tennis court stripes and tidying up shag-piles; but it leaked fuel from the carb and the tank. Two leaks equals too dangerous. And no good for the rough stuff anyway. Another machine had an obsessively rattly blade that frightened me. If that bearing splintered, as it surely would one day, the blade would shoot out from underneath like a supercharged frisbee/ankle-pruner.

The third machine was a narrow rotary cutter. It worked wonderfully, but it took far too long to do the job. And the only way of stopping it was to short it out with a screwdriver. I always felt nervous about doing this.

In the end I scythed the lawns, twice. Hard work, and no velour stripes, but quite a lot of useful hay.

We finally got a new (i.e. very second-hand) motor at the end of October. Without the trace of a smile, the mechanic assured me the machine would carry a six-month guarantee. For a lawnmower. In October.

It turns out to be a cracking motor. I was delighted when it cropped back the wilderness in the Top Triangle, elbow-high hogweed and all. It just roared into it, laid it flat and worried it to shreds. Two passes later, we had a rudimentary Parkland Lawn up there. Very impressive.

So no new £400 machine just yet, thank you.

* * *

I've never been much good with machinery. I put this down to not being allowed a Vincent Black Widow motorbike when I was ten, before the harsh demands of secondary school and O levels kicked in. Had I had this early experience of just fiddling about with motors I'm sure I'd be more confident with them now. As it is, I constantly expect machines not to work, and am childishly delighted when they actually do what they're supposed to. I wonder if this attitude rubs off on the machines? 'Oh, it's old misery-guts again. He wears my bearings down, he really does (*sigh*). I think I'll just rot this morning. No point in starting ...'

One skill I did make a point of picking up before we

came to Wales was welding. The course I went on was in two parts: gas welding and electric. I did really well at the gas work, then broke my ankle and couldn't do the electric part. Guess what gear I needed in Wales ...? Yes ... electric. So I had to teach myself. If you've never tried electric arc welding, and have a masochistic sense of humour, then I recommend it. Otherwise, keep well clear. The essential difficulty is of keeping the electrode rod at a constant height from the workpiece. It doesn't sound too hard, does it? But don't forget that the rod is constantly melting away in order to supply the weld material, so you need to keep lowering it as you weld. Again, not too hard, surely? Well no, not in itself, *but* ...

In order to start the welding process you need to 'strike' the first spark, between the rod and the workpiece. You do this by cunningly bouncing the rod onto the piece, then *immediately* lifting it off the surface to the precise point above it at which to maintain the brilliant jumping blue spark, which is what melts the metal.

If you make the gap too big, the spark is lost.

If the gap is too short, or you aren't quick enough off the mark, the rod literally welds itself to the workpiece. You have to twist and haul and batter it off as best you can, and start again, usually with a now-damaged rod. But it's not impossible.

The fundamental problem is that the blue spark is incredibly bright; so bright that a mechanic friend once ended up with a deeply sunburnt face after a lengthy welding job, with two white orbs where his goggles had been. Goggles, or a face mask, are essential to protect you from blindness.

Now then ... you look through the goggles, and everything is pitch black, isn't it? You dab about hopefully with your rod, trying to find the workpiece. The dog sneezes and has to be locked away, pawing at his nostril. Nothing. Nothing. Noth … AH!! A brilliant blue light. You've found it!

In the next *fraction* of a second you need to lift the rod from the surface perhaps two or three *millimetres*, while constantly lowering it ever so slightly and moving it at a suitably slow and regular pace along the length of the seam you are trying to weld ...

OK smarty boots – *you* try it.

A gift

The East Wing became an immediate godsend. For a start, it was lit from three sides, always an excellent thing. For another, it got light from the southerly directions, while the rest of the house did not. We had become accustomed to living in an interior gloom. Now at last we had a place to actually see some sunshine, and occasionally, a shadow!

It was invaluable as a place of quiet, away from the main sitting-room, which was the motorway from the kids' rooms to the fridge and the outside world and back again. On bad days, when I was too thick to do any real work, I could go in there and read. The light would be good for me too. But what should I read? I was finding that literature was becoming boring. I didn't see the point of it: stories ... to pass the time. But I wasn't interested in 'passing the time'. Quite the contrary.

I needed stronger meat and began to cast about for something that would satisfy.

* * *

We had a wonderful surprise in the summer of 1990. Word had got around the local organic community that the Griffins were having a bit of a time of it, and one evening someone from our old Work Party rang: how would we like a buckshee party at our place? Just for fun?

We were very moved, that people should offer such a thing: to give up half of their weekend for us, with no chance of repayment, and none expected.

Did we have a suitable job for them? What we needed here was one really big long difficult job that ten people could work together on. We pondered, and decided that with a lot of hands we could finally fence the bottom field off from the cwm, to stop our own stock from falling down it, and to stop other people's lost sheep from wandering up the wooded slope and into our fields, eating our grass, possibly spreading disease, and wasting our time trying to find their homes. A mega-fence!

We set a date, and I bought in the stakes and wire and staples we'd need.

On the day, ten people turned up. Then another six. Some of them we'd never met before.

The day was wonderful. My job was the easiest one of driving the tractor round, delivering materials and generally overseeing. Anne pitched in with the proper work of scything brambles back, laying wire, banging posts in and stapling lengths of wire to them. We decided against having one long run of wire, and instead went for half a dozen shorter lengths that could be easily taken down to allow us to haul wood out of the cwm.

At teatime, Anne made vats of tea and trotted out the cakes she'd prepared. Everyone sat in the afternoon sunshine and chatted. Heaven must be something like this, I thought: relaxing after finishing a tough job, well done by a group of friends (and a few complete strangers) as a pure act of kindness. Surely it doesn't get any better than this?

One car at a time, they left to get back for milking and watering and all those other routine jobs at home, and we were left to admire that magnificent new fence. Anne was tired, and I was exhausted, despite having had the easiest job on the site. We went to bed extremely happy with our own lot and with the world at large.

* * *

On the general family front: while Cait was having childish fun roaring round the place, making dens in trees and hedges, Paddy was quietly growing up. He wasn't much interested in the life of the farm. Instead computers had bitten him, along with adolescence. He was an Indoor Youth to the degree that Cait was Outdoor Girl. And when you're a teenager with intellectual and tecchie interests, you just can't wait to get away to the bright lights of the big city. 'Twas ever (rightly) so.

– 20 –
Dumbing Down

The fact that sheep can develop individuality came as a real surprise because mass commercial flocks do tend to be stocked with the brainless bimbos of mythology, who have minimal human contact and thus become herd-bound, a bit like teenagers and street corner gangs.

Their day consists of eating grass, panicking at the drop of a hat (or anything else that comes to hand), then following each other into a stifling corner somewhere, where they can barge and jostle each other till … well, till they just drift off again to eat some more grass, and the circle is completed until the next mass panic attack. Herd-bound.

Our lot are a bit different, firstly because there are only a dozen of them and they thus get a bit more individual attention than a professional sheep farmer can offer, and secondly because as well as a few 'commercial' crosses, we have a couple each of several pure breeds, and individual breeds have individual tendencies.

Cheeky the Jacob, and Bambi and Beauty, the two Leicester Longwools, were the brightest, while the Suffolk crosses remain depressingly dull and conventional. For example, I've never seen a Suffolk try a weed (or 'field herb' as we call them. We're quite strong on field herbs, actually). Cheeky would try anything, even the occasional thistle.

When Porky was little she thought she'd like to play with

the sheep. To start with she was a bit nervous about it and only realised what fun it was when the Suffolks finally twigged that they were being outrageously assaulted and gathered up their skirts and dashed off in all directions, before coalescing into a single unit, squealing and shaking their handbags at their ferociously yapping and tail-wagging assailant. Even the Longwools loped gracefully to one side.

But Cheeky stood her ground …

High Noon. A lone harmonica wails softly as hunter confronts prey, head to head. Eyeball to eyeball, they size each other up, with all the instincts of the ages jostling within, the pattern broken only by the endlessly wafting tail and occasional squeaky yap.

Stillness. A buzzard surfs the sky, watching.

Silence.

A single ball of tumbleweed blows from left to right. A distant mission bell tolls…

It ended, predictably enough, with Porky scuttling for comfort, yelping pitifully and nursing a sheep-skull-shaped bruise to her anatomy and ego. The rest of the flock took note. They were not so easily panicked the next time Porky appeared.

We recently noticed a third difference in mind power. In previous years the Longwools and Cheeky (the Smart Set) discovered how to dig up Jerusalem artichokes when food was scarce in winter. Soon they were all at it, grubbing out tubers with their feet then nibbling up their trophies, mud and all. They *learned*. A good leader is also a good teacher and developer of minds, it seems. Develop the mind, and develop the individual. As with sheep, so with people; even teenage gangs, possibly.

This year is different though. The leaders are no longer with us, and the hoi-polloi are on their own. They panic now when Dylan the dog approaches. Cheeky's heroic stance has not inspired anyone else to stand firm.

And now that we've let them loose in the over-wintering garden, we notice they are not digging artichokes. They've dug artichokes for years, but without Cheeky's leadership and inspiration, they've forgotten how to. I'm reminded of how culture disappeared in Britain when the Romans left. People even forgot how to read and write. ('What have the Jacobs ever done for us?')

Is it a coincidence that the Jacob is a 'primitive' breed? Probably not. Does this mean that the good old days were better for sheep? That evolution has failed to develop their intellect, contrary to general expectation? No ... I don't think so. The reason modern breeds are dumber is that farmers have deliberately bred the brains out of them. Who needs sheep that are clever or curious? They're going to find gaps in hedges, and learn how to swing along under overhanging branches to drop over the fence, then pinch bikes and pedal away to freedom. Brains means trouble.

We've personally seen this dumbing-down process in action. Every now and then one of Ken's horde of norms has a rush of blood to the head, develops some initiative, and breaks into our field. Ken rounds her up with the enthusiastic help of Fly, his pie-eyed and three-quarters crazy collie, and bangs her up in the slammer for a couple of days. If she re-offends, it's a one-way ticket to the abattoir, where she is transformed into lamb (whatever her age) or cattle feed, or muesli, or whatever is currently permissible. So she joins the long list of other brainy ones, the Sheep Who Dared, and never gets to breed, and evolve the stock. The ovine genome project has ground to a halt.

I wonder if we'll ever stop this overwhelming dumbing down when we realise that evolution (i.e. individuation and growing awareness) is for everyone, not just people?

My guess is that yes, we will, but not for quite a while yet.

(If you consider evolution in terms of 'growing awareness' rather than just 'survival mechanisms', things make a lot more sense. At least they do to me. Do they to you?)

* * *

All those sheep inevitably got Anne's creative juices flowing. Wool, wool, wool ... it had to be sheared off them every year, and wool was not fetching any sort of money. What to do with it all? There was only one answer: spin it, then knit it or weave it.

Anne joined a local spinning club, held at the National Woollen Museum, just a couple of miles away in Drefach-Felindre. Elinor showed her the rudiments, first on a spindle and then on a wheel, then lent her a wheel to practise on. Anne picked it up fast (no surprises here) and soon bought her own wheel: a nice little Ashford, of traditional simple design.

No evening from then on was complete without an hour or two of the gentle whirring of the wheel and flyer. I offered to get her a black pointy hat and a cat to match, but was politely rebuffed. The pile of yarn gradually grew larger and more regular. What next?

To knit or to weave? Anne already had a three-foot loom from suburban days, but it was too big and awkward to set up in the sitting room, so she would have to use it in the cold and lonely office. No chance. We enjoyed sitting together in the evenings, so knitting it was. I gawped at the telly, while Anne gawped at the telly *and* spun or knitted.*

After a few nights of clicketing needles I asked, 'Have you any idea what it's going to be yet?' She held up a thick brown mesh. 'Is it a fishing-net? A bath mat?' One part looked a bit tubular and was clearly separate from the rest of it. 'It's a trunk cosy for an okapi or a small elephant. Probably an orphan.'

'It's a jumper for you.'

'Oh ... thanks.'

The yarn was big chunky stuff, spun off a chocolaty Suffolk cross called Bluey. Every now and then:

'Stick your arm out.'

'What for? I'll feel silly.'

'So I can measure this against it.'

I then realised that she was knitting a jumper without a pattern. This was inconceivable to me. How could you work so technically and yet so spontaneously? But that is how Anne approaches everything creative, be it gardening, cooking, or Life itself. You examine the principles, make a mental plan, and go for it. Who needs a pattern or a recipe? Who, indeed?

The jumper was terrific. Very brown, with a thick cable-knit decoration down the middle from neck to navel. I tried it on. Very warm. Anne stepped back to check the fit:

---

* Women seem to be incapable of just sitting quietly. It's my theory that they invented knitting, weaving, pottery, cookery, interior design, cushions, etc., specifically so they'd have something to do in those long winter evenings in the cave while Man the Hunter drank gallons of bee-mottled mead and fell sideways into the fire.

'It looks as though you've been run down by a moped.'

'It's great. Thank you.'

It served well for all work in all weathers for two years, then began to wear thin here and there,* and to look excessively scruffy where it had snagged a thousand times on brambles, hawthorn, blackthorn, sheep's feet, and barbed wire. Then she recycled it into another jumper, with an ounce or two of garish 'Fairground Grey' from a Longwool, added to provide a bit of rainbow gaiety and to cope with my expanding girth.

So, a fashion plate at last. My dreams come true.

From then on, Anne hasn't looked back. She's made a dozen jumpers, several coats and jackets for herself (all of her own design), hats, waistcoats, scarves and even socks. Her first attempts at socks were a bit off-beam (think ... 'foot-bags'), but with a little adaptation they made excellent balaclavas for the dog. She persisted, of course, and eventually got it right. Her tours de force were half a dozen beautiful shawls, subtly multi-coloured, soft as a maiden's prayer, and of striking design.

It took several years of learning and experiment before she made the shawls. After mastering spinning, she learned about washing raw fleeces and how to card them to straighten out the mass of fibres,** then the arts of dyeing, plying, blending, and working with other yarns, from silk and flax, to 'fleece' made from recycled pop bottles, of all things, and (honestly) soyabeans. They're all different to work with, as are fleeces from different sorts of sheep: Southdowns have short

* Especially around the knees. No ... only kidding. No ... *really.* Ow.

** My own contribution was to make a pair of over-heavy carding hands, and a niddy-noddy (used to wind yarn into hanks), which looked nice, but broke.

fine wool with lots of crinkles; Leicester Longwool fleece is long and straighter and with less crimp; Jacobs' wool is surprisingly soft; the various crosses give all sorts of variations of softness, crimp, length, strength and colour. What's more, the wool varies hugely, depending on what part of a sheep it comes from. And you thought wool was wool, eh?

* * *

If you fancy your hand at spinning, the easiest way to start is with a drop spindle. You can make one from a potato and a knitting needle, or a lump of lard and a chicken leg. Or, to be bang up to date, an old CD and a pencil (and a chunk of Plasticene, or similar).

The principle is that you set your little flywheel (for that is what it is) spinning, and gradually tease out your carded wool, allowing the spin to translate into the thread. This brilliant and amazingly simple technology dates back to the stone age and has never been beaten for portability and compact design. Old spindles turn up in just about every domestic archaeological dig. Even the Romans, the original hard-nosed Fascists, had spindles, although one suspects that Julius Caesar knew very little about it. But one may be wrong in this, of course...

> GO AND HACK OMNIA GALLIA INTO TRES PARTES IF YOU MUST, GLUTEUS CALLIPYGUS: SLAUGHTER THEIR WOMEN AND RAPE THEIR CATTLE IF YOU WILL. BUT I! I MUST TO MY SPINNING ... FOR *I* WILL BE REMEMBERED, LONG AFTER *YOU* HAVE CROSSED THE STYX AND BECOME A SHADE ...
>
> FOR I, *JULIUS CAESAR*, AM ABOUT TO INVENT ... ***THE HOMESPUN TROUSER®***.

A spindle takes a bit of getting used to, and beginners are likely to spend some time retrieving it from under the sideboard, or discovering that the wool somehow untwists itself and falls apart at a touch. Or you may get the rate of teasing wrong and end up with a super-slubbed string of small woolly sausages. It's all in the wrist and the rhythm. Gradually your wrist and fingers learn and you let them get on with it. It's a bit like driving a car, I suppose, except that I never lost our Anglia under the sideboard while I was learning.

The next step would be to a spinning wheel. A wheel is really just a spindle turned on its side with the flywheel element (originally supplied by the CD, potato or lump of lard) replaced and automated by a big wooden wheel propelled by a treadle. Another masterpiece of engineering. The key to the machine is the flyer, which loads the new-spun wool onto a bobbin, which has replaced the pencil.

This is the modern wheel. It was preceded by the Great Wheel, which was really just a horizontal spindle driven by, yes, a great wheel. This is the machine that Sleeping Beauty pricked her finger on. No Health and Safety Executive, you see.

Dozy Woman Pricks Herself.

See Feature: Why Are Women So Dozy? p. 12

If you fancy going down the *From Fleece to 'Fleece'* route, you will learn to enjoy the various stages of preparation.

We discovered that it's not just a matter of sitting down behind Ken while he shears the sheep, and teasing off a handful to spin on the spot. For a start, the fleece is usually filthy, well stocked with thorns and bits of herbage, and will always be stiff with greasy lanolin. It needs cleaning.

First fill the bath with scalding water and detergent; sink the fleece into it, stuffed into an old onion sack or fish-net nightie; let it soak while the water cools; lift and turn occasionally, as for a roast; the water will be richly coloured and might make a fine gravy. The brownness of the water is just muck, actually; it's not the colour coming out, as many a beginner has feared. If it fails the gravy test, it will be a pretty good liquid feed for the tomatoes.

Drain, warm rinse, and spin dry, as for washing a parrot. Then dry it. Ideally, you disentangle it a bit and hang it over the Rayburn or spread it on the lawn if it's a fine day (… but not just after mowing it, if the trimmings are still in situ. Can you guess why?)

The fluffy stuff

Another reason for taking up spinning was that Anne thought we might be able to spin, knit and sell some craftwork as a minor value-added diversification, especially as raw wool was nearly worthless, but we hadn't realised the first law of craftwork: it's a dead loss. You can never hope to recoup payment for even a fraction of the time you put into the making of something, even at the rate of the minimum wage, never mind getting a premium that reflects your skill or artistry. That's if you sell direct. Once you get involved with shops you become a sweatshop serf and will have to sit back and watch the shop double your own poor price for the privilege of not displaying it in their window. And of course, they are most unlikely to actually *buy* it off you. It will be 'sale or return'. You do all the work and take all the risk. They take half the money. Or more. Hence the popularity of co-operatives among craftspeople and artists; and market gardeners as well.

We still have a couple of Anne's shawls. She would need to charge £350 a time for them to recoup even a portion of the effort put into them. But somebody somewhere will be sure to think they've got a bargain. And so they will have.

* * *

One evening, Anne started rummaging through her bag of assorted yarns and fished out a wad of fluffy stuff. Immediately, Dylan jumped off the sofa and bounded across the room, eyes wide, tail thrumming, and stuffed his nose deep into the fleece, snorting and panting.

'Has he finally flipped, do you think?'

Anne felt the wool between her fingers and tried to stop the dog from burying his entire face in it. 'It's angora. He can smell the rabbit, I suppose.'

If ever one needed proof that a dog's sense of smell is of a completely different order to our own, well, here it was … because this batch of 'angora wool' actually contained only about ten per cent of angora rabbit fluff. This fluff had been washed several times in detergents, fluffed up, aired endlessly, and dried; then blended as one part in ten into a fleece that would smell, to the same degree, of sheep.

Bam ... straight in. Boy, that dog sure knew his rabbits.

– 21 –
Ditching

Ditching has a rather unsavoury reputation, hasn't it? Redolent of filth and rot and so forth. True, one might occasionally have to remove the best (i.e. worst) part of a sheep from a gully, but by and large, I think ditching is a really enjoyable job. And I'm glad to say that we've never actually had that regrettable problem on our own patch.

SHOCK HORROR!!! Sheep Dies!
Falls in Ditch!!
'I Just Took My Eyes Off Her For A Moment,'
Claims Shepherd.
'Usually So Careful,' Court Hears.

The enjoyment of ditching is to do with instant gratification, a quality of life that rarely crosses the threshold of most smallholders, because we tend to be people who live by such long-term dicta as Lay Firm Foundations … Duty First … Sound Infrastructure is Vital … *Never* Get Into Debt … Any smallholder who doesn't obey these age-old guidelines for a happy life is in for a rough ride, and an early return to suburbia, with 'Failure' hanging invisibly round his neck.

He might find, for example, that he loses all three of his geese to a fox one evening because he is ten minutes late in locking them up; or that the following summer he spends endless hours carrying heavy buckets of water uphill to soothe his gasping lettuces, when, if he'd listened to … ooh, his wife for example, he would have taken the time and trouble to lay a hose line up the hill last winter instead of watching all that snooker.

Ditching is refreshingly immediate. For a start, you know when you have a problem: water comes sloshing through the geraniums/tool shed/bedroom/all three. It cannot be ignored. Secondly, you can see exactly where the problem lies: this bit of ditch, right here, is blocked with silt/twiggery/a double-glazing salesman who unaccountably lost his footing sometime last year, apparently; oh, dear. Thirdly, the solution is easy: clear it out. And fourthly, there is that wonderful experience of seeing the problem instantly rectified. *Instantly!* We even have the visual aid of muddy water gradually giving way to clear, and the tidemark in the bedroom visibly sinking, to leave an attractive dado effect.

The hardest work we ever did on the ditches was when the East Wing extension was going up and we had a huge and clapped-out dumper truck on site.* We started from the top of the drive. Big Ianto shlocked up great scoops of slimy sludge with his JCB, and then gollollopped them into the dumper. When it was suitably and stinkingly full, I then drove, if that's the word, the dumper down the drive, through the yard and right to the slippery edge of the cwm where I discharged the rancid goo.

* Here and there you could still see traces of the original yellow paint on it, dappled like little tiny buttercups across great prairies of dented rust. That should have given me a warning clue as to its general mechanical condition, wouldn't you say?

Nobody told me the brute had no brakes (builders' humour, you know) or that it leaked diesel from every crevice; or that the clutch was amusingly erratic. Every trip to the cwm took me to the edge of the unknown, and a rusty plummeting death. I hated that thing. It was so big it was hinged in the middle to enable it to be manoeuvred at all, and steering from the middle is not a bit like steering a car. Not one bit. Oh yes – the steering mechanism was shot, too. It groaned like The Damned and had between three and eight inches of slack in it at any given time. You could not predict how much. A narrow driveway, a huge dumper, no brakes, duff clutch, weird steering ... I ended up in my own ditch only the once, surprisingly, and had the embarrassment of having Ianto haul me out to massed ranks of builders' jokes. Oh ha-di-ha. I estimate we cleared nearly sixty tons of stuff out of the ditches that day. I tremble to remember.

I've never really thought of ditches as being dangerous, but one incident might easily have made for some light amusement in the coroner's court. At one point, a five-feet-deep storm gully abuts the field full of our neighbour's beautiful Jersey teenagers. They are lovely to look at, but inexperienced: let's face it ... 'dumb'. One afternoon our visitors called me. One of these bimbos was in difficulty. Couldn't get up. Was she ill, or something?

In fact, she was reclining languorously on the edge of the gully and was in immediate danger of falling five feet backwards and downwards into it. End of cow, most likely; also total blockage of ditch.

So what did I do? I scrambled along the gully to try to push her up *from underneath.* Oh folly. All this did was to panic her and she damn near fell on top of me. Only Anne and the visitors heaving at her horns and hooves prevented the headline:

Lovely Cow Falls On Idiot

Why Are Idiots So Stupid?

We Ask A Reality Show Contestant

The Lovely Cow was fine, incidentally.

* * *

We *should* have spread that sixty tons of rich organic matter onto our fields as a general fertiliser, but we couldn't spare the time or face the appalling prospect of driving the brute onto the field and getting stuck, as we surely would have. No doubt we could have hauled it back out with the tractor, but it would all take time and meanwhile we would be paying Ianto more than we could afford for sitting about and not clearing the ditches.

* * *

Should you ever fancy doing a bit of ditching, instead of, say, slobbing around in your pyjamas watching *Match of the Day* with a nice cup of tea and a ham and pickle sandwich, might I recommend the following principles?

1. Pick a proper decent rainy day, so you can see precisely where the blockages are, and exactly where you need to act. This is actually a very interesting task, requiring lots of judgements on how to get the maximum effect for the

minimum of digging, and you just don't notice the rain after a bit. An umbrella can be handy, especially a gaily striped golfing job which will attract the giggly attention of every Jersey heifer for miles. However, the time will come when you need either a third (and, if a little gusty, a fourth, possibly fifth) hand to hold the brollie. A passing jogger might like to help, or you might just resign yourself to getting drenched and the hell with it.

2. Tools: a good stiff rake will haul out most of your blockages, which are likely to be dead grass stems and odd snapped-off twigs rather than solid bodies.

    Garden hoes can be handy, too. A Dutch hoe can break up a small tough clump, and a draw hoe is good for hauling loose gravel about. Pointy shovels are best for cutting out bare earth and clay and pebbles. A small spade sometimes has its uses. But don't even attempt to dig dead claggy weeds out. The rake is your tool here, unless the weed is actually *growing* in the ditch, in which case you'll need a heavy-duty mattock, or preferably a supercharged JCB if the weed is a needle-leafed reed.

    Clearing pipes needs the chimney-cleaning rods. They come with a claw tool that breaks up lumps, and a flat pusher/plunger thing that shoves the rubbish through the pipe and away. However, the attaching thread is so coarse that you must keep twisting the whole assembly as you work, or the tool will surely detach itself. We lost a perfectly good claw like that. It just ... came off while I was reaming out a pipe in a heavy sweep of water. Heaven knows where it ended up. Ireland, probably.

3 Obviously, you start clearing at the bottom and work upwards, otherwise you find yourself working in ever-deepening water whose only desire in the whole world is to first flood your wellies, then sweep you away to Boston.

Each new clearance helps the flow. The process takes you back to childhood holidays at the seaside when the dragged heel of a small wellington would cut out a channel for the water drooling from the cliff face, and direct it so that it would very satisfactorily wipe out that ornate and excessively be-flagged sandcastle built further down the beach by the rich family with the Ford Zephyr and the over-loud voices and pre-crenellated buckets that made castles a doddle, while you only had a plastic plate and a wooden spoon …

I wonder how many professional hydrologists began as embittered and envious children? But I digress.

Once the last gobbet of muck has been prised out of the ditch, it is one of life's joys to watch the water suddenly flush downwards, running from turbid to crystal. It is an elemental act of purification. Suddenly the swampy flooded area is no more. The land separates from the waters, and it is good …

What's more, the sudden release of water at the top will have an acceleratingly flushing effect on the whole length of the ditch, and sweep any remaining rubbish along with it, including several of your tools if you've not taken precautions.

Time now for that long warm shower, then slobbing around in your pyjamas, watching *Match of the Day*, with a nice cup of tea and a ham and pickle sandwich, positively glowing with self-righteousness and the joy of achievement, unlike some others I could mention.

4. An afterthought: ditching work is most effective as far uphill as possible. Thus, in a field flooded by someone else's run-off, a ditch at the top is more useful than one at the bottom as it removes the problem 'at source'. But obviously you might need one at the bottom too; or you might need to lay drains in the field itself; or both. It can be a tricky old business.

   We finally solved our own yard-scouring problem by installing that twelve-inch pipe fifty yards back up the hill, and then by Ken using his mini-digger to do some clever dyke and furrow work in a neighbour's field. This directed huge amounts of field run-off into the big new pipe, so it never got a chance to roar down the lane any more.

5. And finally: it is universally acknowledged to be a crime against the universe not to take a dog with you when ditching. He'll *love* it.

Wood-butchering

While Anne was experimenting with all things woolly, I was mucking about with wood. My previous experience was pretty slim. I was a dab hand at making kindling (a sledgehammer is often more useful than an axe, was one of my discoveries) but that was about it.

We had quite a lot of bits of wood, especially the yards and yards of narrow pine planks we'd taken down from the sitting-room ceiling. What could I use them for? They were straight, and beautifully figured. Too good for firewood or giving to the dog.

As I wander round the house now, I keep coming across odd bits of this planking: backing for a row of coathooks; backing for the blinds in the office; a fifteen-foot row of book-shelves in the Computer Suite; a headboard for a bed; the curvy-carved end pieces for a clothes airer (of the sort you hoist up over the stove).

I also used a yard or two in making a white elephant for Cait. She'd saved up and bought a Korean Stratocaster copy, in shocking pink, and began learning the chords to the Jimi Hendrix Book of Campfire Songs. In a fit of madness I thought that what she needed next would be a 'flight case' (one of those long flat boxes that guitars get hauled across continents in). I had the wood!

I made my own plan (learning from Anne here), and bought in the other bits I needed: fabrics, paint, catches, hinges, etc. For weeks I spent my evenings in the pack-house, wrapped up against the autumn cold, clouting this musical coffin together. To my surprise, it turned out quite well. There were only two snags: the fabric lining ruckled and bubbled and stayed that way, despite re-gluing; and Cait really didn't need a flight case, and especially not one that she could scarcely lift. Bless her, she didn't have the heart to hurl it, very slowly, out of the window. It just got 'put away' and we never spoke of it again.

The chair was more successful. I'd always fancied making a chair. Again, no bought-in design to work from. How hard could it be? Especially as we had a lot of nice straight pine planks and 'tessellated' vinyl on the kitchen floor to help keep the lines and angles nice and true … I even designed it so the seat lifted upwards on a piano hinge, to reveal a useful box, currently filled with shoe polish and brushes that nobody ever uses. No doubt it will come in handy for storing Amstrad floppies one day.

A success then? Ye ... e … s, up to a point. Anne uses it as a computer seat, or sometimes for meditating on. The snag is that it has somewhat Puritan ideas about posture. There is no angling built into the back, you see: too difficult to make, for a first effort. Thus you are forced to sit startlingly upright, like a Victorian Aunt with a carbon-fibre corset and an unmentionable condition.

Another little piece of plank went into making a love spoon for Mum's birthday. This was not carpentry, but carving. Tricky stuff, carving. One slip and you've ruined it, or are looking for your thumb in the fruit bowl. But very rewarding. I made another spoon for Anne out of a six-inch piece of wringing wet and discoloured roofing batten lying where it had been discarded the year before in the yard. So forlorn; wet, unused; unuseful … scarcely worth the effort of picking it up for burning. I was delighted to rescue it and make a spoon from it, with a carved flower in the handle.

Perhaps the most ambitious thing I attempted was making a wooden chain. This time I did use a pattern, or at least I followed the instructions from a book. It turned out to be remarkably easy. I'm no longer impressed by wooden chains. But I am deeply impressed with the Grinling Gibbonses of this world. Perhaps you need to try it to appreciate the superlative skills involved. I would say that a master woodcarver needs all the skill of the sculptor, and then some more, as he needs to be paying constant attention to the grain, which a sculptor does not.

The most difficult things I made were ladles. The outer surface is simple enough, but getting the inner bowl smooth is very hard indeed. At least, I found it so. Perhaps you know otherwise.

I made three of these. Two I gave away. They were my later efforts, carved from sycamore, with big deep useful bowls to them. The other one, my first effort, is hidden discreetly away in a kitchen drawer. Like the flight box, it was not quite just the job. The bowl was long, rather than circular, which meant it didn't function quite right, and the decorated handle was not good enough to rescue it. Anne thought it looked like a Dutch clog with a stick of barley sugar cobbled onto it.

I believe I've already mentioned the carding hands and niddy-noddy. Ahem.

* * *

Before we'd properly taken it in, Patrick had suddenly grown into a young man and was ready to spread his wings. One day he was doing exams, as ever; next day collecting his results; then he'd gone, off to university in Birmingham to study things technical.

It must have been a sort of teenage heaven for him, being with lots of people of his own age, with similar growing-up interests. Out in the sticks, you have a much smaller circle to pick your friends from, and then there's the problem of transport. Everyone lives miles apart, and events, such as they are, happen even more miles apart. I think it would be true to say that a 'world-citizen' teenager was not spoilt for entertainment choice in our district. Folk music has limited appeal, as do gigs by Billy Turnip and his endlessly vibrating Hammond organ, even if accompanied by Doris with a cordless mike turned up to twelve, fretting your eardrums and setting up standing waves in your beer in a public bar thirteen feet by twenty.

Big-name groups never come anywhere near West Wales, and local bands tend to be, well ... local. Limited talent, bolstered by fifteen FX pedals and a belief that if no one can hear the words you must be doing something significant. One of these bands was called the J Arthurs. At least they had a sense of humour.

As we lived well below the official poverty line, we couldn't really afford to taxi Paddy everywhere in the Volvo. He knew this, and never complained, but it must have been a serious problem to him in those crucial years. Now he could try the big city.

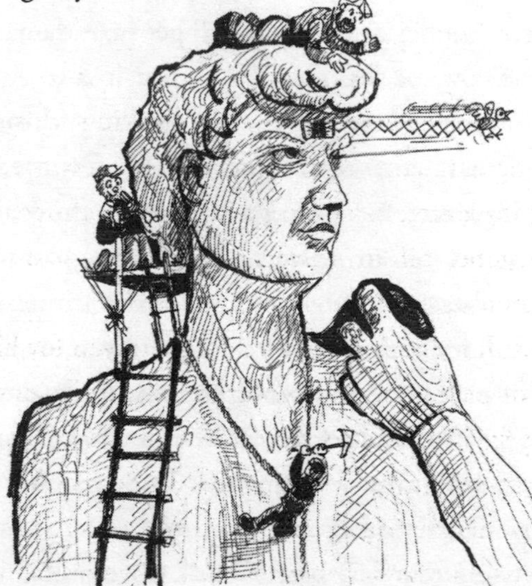

– 22 –

The Marrow

'Give the cow a marrow, too, while you're out there.'

'Give the cow a marrow'? What could be simpler? Fetch Marrow A and present it to Cow B. Two minutes, tops.

What is it with this universe? Why does the simplest task feel itself obligated to expand into a full Three Act Opera Bouffe? It wasn't like this in the last universe. Things properly organised there. Mutter mutter ...

The marrows live out their remaining days in relative peace and quiet on a length of staging in Polytunnel 1, along with the pumpkins and squashes. When frosts threaten they are lightly draped with horticultural fleece and left to take their chances. Sooner or later one will get frost damaged, and thus become cow-fodder. This is no great loss to Anne who reckons you expend more energy hacking through the marrow's Kevlar skin than you recoup from eating its contents. I, however, being thrifty by nature, think it no bad thing to have the reassuring bulk of a 12lb marrow in reserve for when the blizzards brought on by global warming coincide with the invasion by the revivified Red Army, and all the oats and potatoes have gone, and Anne approaches, wiping a tear-sodden cheek. 'Oh, husband, dear! Whatever shall become of us?' she will whimper, wringing her poor careworn hands through the mangle of her emotions.

Then will I murmur sagely 'Well, wife: we still have ...

THE MARROW, safe in the polytunnel. Good thing we kept it, eh? And didn't *give it to the cow*, or anything?'

So … we have three marrows remaining.

I clamber over the sheep hurdle which serves as a door to the tunnel, and prod the first one.

It is firm, yet soft, like an armour-plated éclair, and gives only a dull 'tub-tub-tub' when you tap it. Oh dear. Frosted. That's for the cow, then. But the second one rings true as a gong when tapped. Eatable! So, with a 12lb marrow under each arm I lift my first leg back over the hurdle. But I am now nearly two stone heavier, and there are two inches of mud in the tunnel entrance area. Stuck: rather painfully straddling the hurdle. The dog looks on, sardonically. Then he drops his stick and hurtles off like a scalded ferret. He's seen a rabbit. That means chasing it at the full gallop into the cwm, with the risk of surfacing eventually in someone else's field where he'll be shot for sheep-worrying. Poor soul: he *looks* like a sheep-worrier.

Meanwhile, I'm still stranded and straddled, calling 'Please return, you foolish dog!' or similar, in a shrill contralto. I lurch forwards to shift my centre of gravity enough to haul my back foot over the hurdle … and drop the soft marrow. It lands on my front foot with a dull 'thwuk', like a wet sheep bouncing off a roof, and spills its orange guts. My back wellie remains in the tunnel, stuck fast.

But I'm out. And the good marrow is still intact. More yelling brings the dog back, grinning wildly. He's now not remotely interested in chasing his stick, and won't be for another hour. Instead, I get a blow by blow account of the whole puerile episode, over and over again. 'Yes, well ... now you're back. So come on, help me take the marrow to April.'

No dice. He lies down full length in the wet grass, like a cheap sphinx, beaming bonhomie and the spirit of adventure while I salvage my other wellie.

I carefully pick up most of the shattered marrow and set off for April's shed. It's got an old house door to it, complete with letterbox, and next to it a modest glassless window, through which we post her hay into an old wheelbarrow. As I approach, I can hear her crunching at a wisp of hay (not the best we've ever had, it must be said). Her eyes light up and she leans forward, tongue curling towards the dripping marrow. 'Steady girl!' I advise, and offer her a strip of dripping orange flesh. It's gone in a flash. Another piece. Another. I can't keep up, but don't want to give her the whole thing at once for fear of her choking on it.

My penknife is a little Pakistani job. Brass, with wood inlay, and a stainless steel blade. Used for skinning yetis, I was assured at the time of purchase. I keep it very sharp, and am conscious that the safety-lock is useless. So I'm careful.

Gosh, that Kevlar is tough. The blade keeps on getting stuck, or jack-knifing – yes, that's the word for it – and all the time there's a foot-long prehensile tongue trying to get a purchase on anything that isn't twiggy hay. A penknife would make a nice change. I can just about keep up, hacking off two-inch rings, then roughly chopping them into five or six chunks. She downs them in one, like a starving gourmet who has stumbled upon a barrel of oysters.

Almost done. My gloves are wringing wet and slippery, so I remove them and immediately get a needle-sharp sliver of hay down my thumbnail. I squeal out loud and jerk my hand back, narrowly avoiding sacrificing the cow with the dripping blade, and bash my watch on the windowsill, scratching the lens and jellifying the works inside.

At that precise moment, and with a sudden lurch, April climbs into the wheelbarrow, to snatch the nub-end of the dripping marrow from out of my hands. Only the front legs in fact, but she's wobbling precariously, and vets cost a fortune. I beat a retreat.

'Did you remember to give April her marrow?'

'Yes.'

'You've been a long time. What kept you?'

'Nothing.'

'Bleeding again? Here, let me ...'

* * *

We only once grew marrows as a proper crop. Usually, they turned up by accident, as over-excited courgettes, somehow overlooked under the big floppy leaves.

We also once tried growing some exotic gourds, on the principle that while marrows are huge and heavy and of very little value, gourds are small, light, and being of only decorative value, are therefore worth more per ounce than anything nourishing could ever hope to be.*

We crammed eight plants onto twenty square feet of spare land, and ended up with a couple of dozen small fruit, each a few inches long, of various and beautiful colouring and design: cricketball-sized smooth white globes; beige light

---

* People in Britain *will* not pay for quality food. Alright, alright ... yes, I know this is an over-simplification and that you personally, dear reader, will go to a lot of trouble to source wholesome food; but having said all that, my original whinge is *still* true. By and large.

bulbs; green and white striped ovoids; ditto but with a bottleneck and, unbelievably, a yellow top, like the capsule on a fine Sauternes; and items the size, shape and colour of your standard orange, but with a crusty scrambled sort of skin that prevented them from ever being mistaken for a mini-Jaffa.

Try as I might to find a way of eating them, there was no way round the fact that they *were* strictly decorative – the skins were actually porcelain-hard shells, with no juicy watermelon flesh inside, or indeed any pulp at all. Just a handful of skinny seeds. So what was the decorated shell all about? Why the 'promise', if no reward? Nature, eh?

The thing to do with them is to scatter them casually into a basket in the kitchen and just enjoy their extravagant colouring, or I guess you could try flogging them as rattles at the Practical Voodoo (Beginners) class in your local college. But definitely not edible.

We flogged a couple of dozen at the WI market. I wanted to sell them along with a recipe for Gourd à la Buttered Toast until Anne reminded me of the Trade Descriptions Act.

A success? Yes, but almost certainly not repeatable, because not consumable. No return orders were likely.

One day we were surprised when a couple of people with very long hair turned up at the farm, out of the blue, just to buy these gourds they'd heard of. They cleared out the rest of our stock. Very nice. I hope they enjoyed them.

* * *

The gourd/marrow issue is a good example of how a grower needs to keep a close eye on such boring things as Income per Foot-Row.

A gardener just grows a bit of what he fancies, but a professional grower must be much more careful. Perhaps he can sell all the pumpkins he can grow (unlikely, actually) but they don't fetch much for the room they take up. Much better to grow carrots. On the ten feet of bed-space taken by one sprawling mega-pumpkin plant you could grow three hundred tasty carrots. Or a hundred and fifty delicious onions.

As a rough example of what I mean: on a one yard (or metre) square, you might grow either one marrow plant, or twenty-five Little Gem lettuces. Your crop (God, rabbits, pigeons, and caterpillars willing) will be two or three marrows or twenty-ish lettuces.

Each marrow will wholesale for, say, 40p. Total = £1.20.

The Gems will fetch, say, 20p each. Total = £4.00+.

These quantities and prices are approximate, but the ratio is about right. Clearly, it's a much better idea to grow the lettuces than the marrows, all things being equal (which they never are).

* * *

It follows that it's vital for a grower to find the optimum spacing between his plants. Wasted space means reduced income; and income is slender enough at the best of times.

Our little co-operative exchanged notes on how we were growing our crops, especially our courgettes. We worked by trial and error, always obeying the organic principle of 'Feed the soil, and the soil will feed the plants', and experimented with our depth of ploughing, spacing, watering, weeding, etc., until we found something that worked for us. We knew we were all getting good at it, each in our own way. We also knew we couldn't follow any sort of imposed masterplan, as every holding and every field is different. Following Absolute Rules would be a recipe for disaster.

Nevertheless, the Secretary thought it might be interesting to hear what a (non-organic) government expert had to say on the matter, so he invited The Man from the Ministry to come and speak to us. We knew he would be talking only from theory, and chemical-farming theory at that, but maybe we could learn a trick or two from him?

If it wasn't so pathetic it would have been hilarious: The Expert told us as a *fact* that we weren't ploughing nearly deeply enough, by a factor of at least one hundred per cent; that on average we were spacing the plants twice as densely as The Expert Recommendation; that we did not irrigate nearly enough; and that my soil, at least, was far too acid to even *think* of growing courgettes on (azaleas and buttercups, possibly, but quite hopeless for *Cucurbita*).

Despite apparently doing everything wrong by a large margin, every one of us 'umble peasants was regularly cropping courgettes at twice, or more than twice, the maximum rate per acre The Expert said we could ever possibly hope to achieve.

I was so miffed at this pompous ignorance, and the harm it would be doing to other growers who were thinking perhaps of converting to organic methods, that I wrote an article for a trade magazine, pointing out that this 'Government Expert' was nothing of the kind. They printed it, to my surprise.

The following week, The Expert riposted, angrily. His single line of defence was that I hadn't actually been present at the meeting. He didn't deny any of the points I'd made, however.

Since then I've been very sceptical about Experts. I rather like that old etymology:

> ***Expert:*** *one who says he knows better than you do. (Derived from 'ex-' a has-been; and 'spurt', a drip under pressure.)*

* * *

Another crop that takes up a lot of room is purple sprouting broccoli. Thus it is not much use as a commercial crop for a small grower. But it's invaluable for home consumption, as it supplies its vitamin-rich heads over winter and right through into the Hungry Gap.

> *So what is this Hungry Gap? I bet not one person in a thousand in modern Supermarket McBritain plc has a clue what it means. But even just a century back every person in the country would have known, even the city slickers of the day, because we were all a lot more aware of where our food came from in those days, and we all knew how close to the edge we actually live: from one harvest to the next. That's* it. *That's a fact that hasn't changed from one century to the next, despite the Internet and Hi-Def TV, and sixteen shades of Bright Blonde Bimbette highlighting mousse, because I'm worth it.*
>
> *It's a fact of Nature that crops grow in the summer and are harvested in the autumn. They are eaten over winter (if you can store them safe from frost and vermin), and run out, or finally rot, in the spring, from*

March to May give or take, just precisely when there is nothing else cropping, and thus nothing else to eat. That's the Hungry Gap. Only our freezer (and currently cheap oil, for transport) keeps us from its full horrors (for the moment).

If you like broccoli and would like to try something a little more exotic-looking, take a look at 'romanesco'. The flowering heads form, wait for it … *spiral cones.* To some of us, words like Fibonacci and fractals might come to mind. To others of us 'Mmm … tasty' might do so.

Paddy moves on

Paddy leaving the nest meant an appreciable drop in our income, as his Child Benefit money stopped overnight.

Although we were frugal in all things, we knew how close to the breadline we were sailing. Child Benefit was not pin money to us. It was an essential part of the economy. We bought our clothes largely from charity shops and grew to realise that Oxfam is actually rather expensive. We never wasted a scrap of food. How could we? We knew the full cost of how it was produced. Our minimal leftovers went into stew or soup for the next day; fatty bits from the meat went to an appreciative dog; peelings (not many, as we scrub our veg rather than throw half of them away) went to the cow and the sheep; stumps and stalks, and what little the stock wouldn't eat, went onto the compost heap.

Even though we managed to increase our income, little by little, year on year, we were still struggling to make ends meet, and all the while the ME was around we could never make any proper headway. My 'work-value' was cut by at least a half, and that lost fifty per cent represented almost all of our 'potential disposable income', i.e. money left for development (and luxuries like a cup of tea and egg and chips on a rare day out), after paying essential bills. By any government standard we were so far below the poverty level as to be virtually invisible.

But ... we are most grateful to note ... not *completely* invisible. Society, in the form of State Benefits, kept us afloat when we most needed it. I didn't know whether to laugh, cry, or scream when Mrs Thatcher told us there was 'no such thing as society'. Who, I wondered, did she think paid her wages?

It is a key sign of civilisation (like sticking to the rules of the road) that our citizens agree, via the ballot box, to chip in to support the unfortunate, and I feel privileged to belong to such a society, when billions of others do not. And now that I may once again become a taxpayer, I do not begrudge a penny of it.

* * *

One fund we drew on for a twelve-month period was the Enterprise Allowance. Some say this allowance was a cynical ploy to get names off the unemployment list; others say it was a genuine attempt to help people set up their own small business. Whatever the truth of it, I can now proudly boast to have once been the only nationalised garlic grower in the country. A man in a smart car came to interview us. Then I had to attend a course, with lots of colourful slides and spreadsheets, in which it was carefully explained to us how we should keep track of our incoming and outgoing stock and how to project our profits five years ahead. 'How does this apply to growing garlic?' I asked. The man in the tie paused. 'Well, probably not very much, I suppose.'

True. True.

– 23 –
Flaming Fire

Some of us have the Gift of Fire. Some do not.

It's not that I'm particularly afraid of a good blaze. I can stand and stare at a bonfire until my eyeballs squeak and my ears start peeling as well as the next man, woman, or indeed, dog. It's just that the skills of getting one going and tending it successfully seem to elude me. I'd have made a rotten vestal virgin.

Anne (She Who Understands Things) most certainly *does* have the gift. If there is the tiniest glow in the firebox of the stove when she comes down in the morning, she will coax it into a blazing furnace in five minutes flat. A scrunch or two of Dad's *Daily Telegraph*, a wisp or two of twig, an apparently careless juggling of doors, flaps and dampers, and the hobs of hell spring up before her, crying out in a flamey sort of way for coals or martyrs.

But when *I* try to do it, it's a different story. Be it the merest glow, just discernable in the stygian reaches of the firepan, or a warm and cheery gleam that is positively begging to be swept into hearty flame ... either way, I seem to have the gift of killing it stone dead.

But why? It seems to me that I do the same as Anne does. And I remember that as a child it was my job to light the fire in the sitting-room grate, using only yesterday's *Liverpool Echo* and half a bucket of nutty slack. And I was very good at it, I seem to recall. So what's going wrong here?

I riddle the grate out with the same over-short poker that Anne uses; I scrunch the paper just so, making sure that there's a good long tongue protruding enough to get a match to eventually; I add small twigs, then large twigs (sun-dried to Delia's own recipe), and top them with a small joint of ash; put the kettle on, and by the time I've cranked up the Lamentations of Jeremiah on the CD-player ready for my morning aerobics … the glow has gone. The firebox contains only inky blackness, and that strange feeling of absence, felt otherwise only at funerals and in Chinese takeaways in mid-afternoon. Where am I going wrong?

Even if I re-lay the whole thing, getting filthy and sooted up to at least one armpit, I'm still not assured of success. As often as not, the paper flares beautifully, and burns back as far as the crisp and tindery twigs and GOES OUT *again*. In fact, the last time but one I tried to re-lay the fire, not only did the paper go out *again*, but I managed to burn my thumb with the lighter flame. (Anne always uses long matches. Oh, ha ha.)

After each sudden extinction I find myself gazing into the black hole in slack-jawed wonder, wondering, for the *n*th time, how do public buildings ever burn down? And how do ships, wallowing in billions of tons of water, catch fire? And how did Hitler ever pull off the Reichstag job? Did someone spend hours fumbling with a box of matches in the dark: 'Ooh … nearly got it that time. Hang on … Can I just borrow your hat to fan it with, *Obergruppenmeisterbahnhofffuhrer*?'

No doubt proper arsonists use accelerants, but for heaven's sake, I'm only trying to light the stove. I don't think a pint of petrol is appropriate here. And, if the truth be told, I'm the only person I know who has put out a bonfire by pouring paraffin on it. True, it was only a little fire, and true, it was a bit damp, and true, it was having a bit of a problem with the wind; but I felt the cold shade of Nero shaking its head and raising its eyes to heaven behind my back as the generous splash of kerosene flooded the wan flame into extinction. I didn't care to tell Anne at the time. A man has his pride.

At the other extreme of the spectrum was the time when John the WWOOFer and I were trying to convert a patch of thorny waste from Rabbit Hilton to decent pasture. We piled the brambles high and lit the newspaper beneath it. The fire took light instantly, cracklingly, and vertically flamingly … The heat was remarkable. Then a sudden breeze got up, and the huge ball of blazing brambles rolled off down the hill towards the cwm and the Teifi like a red-hot sphere of infernal tumbleweed. I had visions of it rocketing out to sea leaving a trail of broiled sewin in its wake. 'What's for tea tonight, Megan?' 'Three tons of boiled trout.' 'Oh, I'll nip out for the chips then, shall I?'

* * *

The Tirolia is a very efficient stove, with a nice big oven, suitable for brewing up cauldrons of stew/stewp/stoup/soup, reviving sickly lambs, and for warming up frozen toes after a day in the mud.* It also provides our hot water, and heats the whole house, thick walls and all, as long as we are careful about keeping doors closed and turning off radiators in halls. In fact, it keeps the place so warm that we really don't need the big log burner in the sitting room. We've not lit it for years,

* Yes, even if wearing wellies.

even at Christmas: just too hot. (Mind you, that room is insulated rather better than a shedful of sumo wrestlers. We made sure of that when we did the place up.)

So yes, the Tirolia is an excellent tool, but there is a design problem. Well, not exactly a problem, but a *feature.* Instead of the chimney releasing the smoke straight up and away, the stove has a complex flue system that leads the hot exhaust around the oven, in order to extract as much heat as possible. All very good, of course, but the snag is that a cooled flue does not draw like a hot one does. And a flue that won't draw means trouble.

The other problem with the Tirolia was its position when we moved in. It had been built onto the stupidest wall in the house, both from the thermal efficiency point of view, and from the why-won't-the-sodding-thing-draw point of view. On a cold still night, it would flood its haze of poisons into the kitchen rather than sending them up the spout. It's a miracle that Porky survived the winters.

Even Anne had trouble getting the brute to light if the weather was anything less than frisky. At the touch of a match, the paper would catch, and the twiggery would begin to crackle, but then streams and ropes of smoke would start to pour from under the lid of the hotplate, then ooze and drain from every joint in the edifice, to roll and ripple across the floor like purgatorial ectoplasm, reducing spiders, woodlice, and the dog to emphysemic and hack-ridden wrecks.*

I tried cutting back trees and odd shrubs to improve the draught, but nothing worked. The only moderately sure way

* A roomful of hacking woodlice has to be heard to be believed.

of getting a proper fire was to stuff a sheet of blazing *Telegraph* into the pop-hole hatch in the flue, to encourage the evil grey smoke to follow its example. The pop-hole, however, was on the *outside* of the wall, at the end of a muddy and shrub-shrouded alley, jammed with rampant rhododendron and Japanese quince and camellia, which had all been planted far too close together ten years before we took possession. And, of course, the inevitable fringe of tall droopy nettles, whose only aim in life was to sting you on the back of the neck, *wetly*.

Not much fun on a cold and drippy morning. Once you'd forced your way round to the pop-hole your paper was already damp, and then your matches would blow out, one after the other.

Stupid Hippie Can't Even Light A Stove!!

See Our 'Stupid Hippie' Mug Offer On p. 3.

He Has Sooty Arms, Burned Fingers

And a Pile of Dead Matches.

A Hilarious Gift!

We eventually got so sick of the whole pantomime that we asked James to shift the stove into the kitchen where it should have been put in the first place: centrally, on an internal wall, with a chimney next to it, pop-hole and all.

He laid a dais of quarry tiles, rescued during the Improvement Grant reflooring job, then he and Chris heaved and hauled the heavy steel carcass onto it.

Next, James assembled a new chimney from blocks of volcanic ash. He continued right up to and through a tiny hole in the ceiling and roof. Impressive work.

The chimney stack was now a good eight feet higher than it used to be, and in a much better position to catch the breeze. All it now needed was a pot.

We'd tried a couple of fancy tin things previously. One was called a 'lobster-tail', presumably because it was about as much use as. The other one simply burned through and rusted away.

James said we needed 'a Marconi', which turned out to be a big tall conical pot-within-a-pot, laced with vents on the under rim of the outer pot. Something to do with venturis, I wouldn't be surprised.

Anyway, it's wonderful. The stove became immediately easier to light, for one of us if not for both.

On a cold still day, or if the stove hasn't been used for a while, we still need to do the thing with the blazing *Telegraph*, but now the flue pop-hole is right next to the stove. Bliss.*

But nothing's easy is it? Especially if you cut corners because you are in a hurry. Just the once, I didn't bother with the flaming *Telegraph*. It was blowing a bit of a storm outside, so I thought the draught would be plenty adequate without it. And so it would have been …

However … as soon as the match hit the tongue of the paper, the eye of the storm zeroed in and settled perpendicularly overhead. An immediate dead calm. No draught. You could hear the gods giggling. 'Nice one, Zeus. That'll smoke him out' … 'That'll kipper him good and proper' ... 'Bet he hacks his lungs to shreds.'

And the Great Ooze began …

… from under the hotplate, then squeezing out of every

* Bliss? 'The removal of pain'.

crevice, curling and tumbling like Niagara Falls. And once it's started, there's no stopping it. It's too late to try heating the chimney, even if you can hold your breath long enough to find the low-level pop-hole in the calf-deep smoke.

Perhaps predictably, as soon as the kitchen was stuffed full of grey smooze, the eye winked, and the storm rollicked off again, and I had to immediately go out in it to unblock the pipes.

When I came back I couldn't get a cup of tea because I couldn't even *see* the kettle, so I put on an extra jumper and settled down in the sitting room to read and get my cool back. Then the power went down.

An hour later it flickered back on. I dashed into the kitchen with an electric fan. Back out again to get a lungful of oxygen. Then back in again ... and a dash through to the back door, to hurl it wide to let in some air, notwithstanding the storm, rain and all the rest of it. Back out. Deep breath. Back in to open a window to complete the through draught. *Jammed.* Choking retreat.

Back to reading. Power down again. Cold. Fed up. Cold and fed up. Cold, fed up, and grumpy enough to write to *The Times.* But too cold.

That'll teach me to cut corners with that stove.

* * *

It's getting very old now. Several knobs and handles have dropped off, and the damper and thermostat have packed up completely. One of the doors doesn't really want to stay shut. The stove still works, but it's clearly past its best. A month ago we realised it had started drawing badly again, and was smelling a bit smoky, too. We eventually discovered a small hole in the steel backplate. It was easily repaired by leaning an old oven tray up against the hole, and wedging it tight with half a fence post tapped home with a big hammer.

Days numbered then … but where do we find the £5,000+ for a new stove? Almost a year's income …

* * *

When James shifted the stove, we reduced the old 'chimney' to a three-foot stump, so it wouldn't be tempted to fall over.

One winter's morning, at 2am, we were woken by the alarming sound of running water. The hot tank was emptying, and making a lot of fuss about it, and the cold tank was desperately trying to top it up. A leak!!

Up. Coat. Torch. Hunt …

More hunt …

More …

'Well, where the hell is it coming from? I've looked *every*where.'

'What about the loft in the extension?'

'I've been up there.'

'No leak?'

'No. Not one. Two dead zebras and a box of umbrellas. But no leak.'

'Look again?'

'Grr …'

This went on for the rest of the night. While I was climbing into lofts and poking under cupboards, Anne was turning off every stopcock she could find, because this would surely stop the leak, somehow, eventually. Wouldn't it?

I have no faith in logic when it comes to plumbing. To me, it's as much a Black Art as electronics and cooking. All I knew was that water shooting out of the hot tank while the stove was still hot might well mean either an explosion of the tank, or the total opposite: an almighty scrunching shriek as the cylinder implodes and wraps itself round the element, like one of those shrink-wrapped cucumbers that are impossible to open. Trouble, either way.

The next morning, bleary, we did the Sherlock Holmes bit: opium (well, *coffee*); violin (mimed); checklist; careful thinking – and having checked and rejected *everything* else, only the unthinkable was left: it *must* be in that blasted chimney stump, outside.

And yes, that's where the leak was. Why there was a hot water pipe in the chimney stump we have no idea. We can only assume it was born of the same flight of whimsy that led a previous owner to run central heating pipes round the *outside* of the kitchen.

So that was the Last Laugh from that ghastly chimney/stove set-up. We found the pipe, capped it and cursed it, and it's been as good as gold since.

Diversification

Diversification ... a word that is on every farmer's mind these days. As primary produce loses value, due to the pressure from supermarkets and imports, what can he do to make an honest living? Roughly speaking, he can either start producing something different, or somehow 'add value' to his original crop. He might branch out into guinea fowl, or B&B, for example, or try his hand at making his own smoked ham instead of just selling his pigs for slaughter. He will still have huge problems, not least from government regulations,* but he does at least have the Internet on his side these days. 'Niche' products *can* earn their keep.

Ken next door tried two quite different ideas: growing worms for land reclamation projects, and turning part of his wooded cwm into a trials bike course. Each had its strengths, and thrived for a while, but both ultimately failed, for reasons beyond his control. The eternal problem.

We were always on the lookout for some way of bringing in a bit more cash, particularly as I couldn't do much hard work any more. Anne investigated selling woolly things, and did a course in proofreading, while I tried sketch-writing for television shows, and writing awful verses for greetings cards. None of these ventures came to anything. I even gritted my teeth and tried a bit of tutoring, but it wasn't very cost effective, once travel, time and petrol was deducted from the fee.

We really did *need* a bit extra, though. Moving to the sticks had meant a seventy-five per cent drop in income, which we had expected and thought we could cope with, as we would have no mortgage interest bleeding us dry, but the illness had made a crucial difference.

Anne eventually got some part-time work at the Citizens' Advice Bureau in Carmarthen. Analysing and organising has always been her flair (apart from cooking, lighting stoves,

* Which effectively destroyed our plans of growing tons of organic bean sprouts in a refrigerated lorry container.

designing, etc.) and she really enjoyed the break. The money wasn't much, but it was to us.

Meanwhile, I thought I might earn a few quid taking wedding photographs. I now quake at the thought: such responsibility, all hanging on a couple of rather old amateur cameras.

My only previous experience had been in Nottingham, years previously. The camera had jammed after the first snap, and I had to leave the happy couple on the steps of the Registry Office while I nipped into the Art College to use their dark room. That should have been a warning, shouldn't it?

But fortune favours the foolish, and by and large, the jobs went well. After all, I did offer a very nice cheap and cheerful service. Instead of the hundreds of pounds most wedding photographers ask for, I charged a basic £1 a picture, minimum sixty pictures, with the option of seventy-two (i.e. two rolls of film). After a few sherries at the reception, most people extended to the full seventy-two.

When the prints came back from Boots, or whoever else was offering a budget service that week, I drove them round to the newlyweds, who took orders from their family and friends for reprints and enlargements. It was a nice simple system. They got a nice mix of formal and informal pictures at an affordable price, and I got a decent return for the hours put in; *and* a free dinner. The people were all very nice, except for one Mum who wasn't best pleased that I had somehow managed not to take a special pic of her in her new hat. She did not re-order a single photo. I guess I asked for that one.

But overall, our best 'diversification', if you can call it that, came to us via HDRA (see *Introduction*). One of

HDRA's purposes is to spread the use of Russian comfrey as a compost and soil-conditioning plant, and they sold plantlets via their magazine. As the trade grew, they decided to outsource. Could we grow plants for them to sell on? You bet.

All you needed to do was lift a parent plant, and chop one-inch chunks off all its carroty roots. Then you drilled a shallow furrow and dropped the bits of root into it, three inches apart. Virtually every one of them turned into a new plant, and *quickly*. They cropped very early in the season too. We grew them by the thousand.

These comfrey sales filled our Hungry Gaps for six years, until HDRA decided they'd rather buy from someone else. This was entirely my own fault. I knew that a shred of root a quarter-inch thick would grow a little leaf which would eventually grow on into a thundering monster, five feet high. What I hadn't properly accounted for was that a tiny plantlet didn't have the stamina a bigger one would have. This mattered, because it would have to survive

(a) being posted two hundred miles to HDRA, in Coventry, where
(b) it might have to hang about for a week or more before
(c) being posted off to Cornwall or Shetland.

A lack of imagination on my part. Comfrey sales stopped in 1991. But that was OK. We'd get by somehow.

– 24 –
Archimedes Rediscovered

CAUTION: THIS SECTION IS HIGHLY TECHNICAL, INVOLVING AS IT DOES SUCH THINGS AS NAILS AND BRICKS, NOT TO MENTION CROWBARS. THOSE OF A NAMBY-PAMBY OR ARTY DISPOSITION SHOULD NOT EVEN *CONSIDER* READING IT AND SHOULD SKIP STRAIGHT ON TO CHAPTER 49: 'WE DESIGN OUR FIRST SATELLITE'.

Many years ago I reported that I had spontaneously reinvented the lever when erecting a polytunnel, thus saving us a lot of time, effort, and foul language when tensioning the skin. However, I didn't go into details.

Somebody wrote to me to ask how exactly to use this leverage stuff, on account of he'd heard good things about it, and as he had this contract for knocking up a bit of a domething in Greenwich, he could use a few little tips and dodges on time- and cost-cutting. I duly sent Mr Trotter of Peckham my arcane secrets and wished him well. That was some time ago, and I often wonder how he got on with the job and whether it turned out to be the Nice Little Earner he predicted.

It has since struck me that a few other potential tunnellers might not yet have stumbled on the fact that the humble lever can turn rupture into rapture. So, for everyone out there who thinks polytunnels look too much like hard work, here goes:*

(SfaS) for general instructions on how to put a tunnel up.

Leverage and how to use it

Each of the curved ribs of our tunnel has two nail holes (about 3in apart) which go right through both sides of the rib at one end. This end is meant to drop into the ground-tube. Our ground-tubes have no nail holes in. I assume other tunnels are similar. If yours isn't, tough. You'll have to improvise. But honestly, it won't be difficult. This isn't rocket surgery.

1. Pass a 4in (or bigger, if possible) nail through the uppermost holes in a rib. Repeat for all ribs. (If there are any spare ribs, don't worry: one of the dogs below will help you out.)
2. Drop the rib into a ground tube. The nail will now be resting across the top edge of the ground tube, while the second pair of holes will be down inside the tube. Repeat for all ribs. Attach ribs to spine. Attach door-frames. Now is a good time to plumb in the jacuzzi, if you've gone for the luxury model.
3. *Sw-e-e-p* the plastic skin gracefully over the spine and ribs (a little light Haydn or Mozart might help here; gingham dirndl optional) and roughly entrench it on both sides as per normal.

---

* Please note that I'm well aware that one diagram is worth a thousand words. However, you have not seen my attempts at drawing a football, never mind a complex technical diagram involving nails and holes. Be grateful I haven't attempted one here.

4. Affix the skin to the doorframes, using an Uzi-style stapler, but *not* a nail-gun.
5. Collect:
   - a brick or two, or some wood blocks;
   - some thin scrap wood to act as shimming material to build up the bricks or blocks by an inch or so, if necessary (see (7) below);
   - a long strong lever, preferably a 4–5ft steel crowbar, spike or tube; a 4in nail;
   - a can of your favourite long cool tipple (two cans are acceptable on uncommonly hot afternoons, when the sky is blue ... odd fluffy cloud; birdsong perhaps ... sheep line-dancing in the distance ... surely the tunnel can wait?) Sorry about that. Back to the job ...
6. Place the brick(s)/block(s) as close as reasonably possible to the point where the nail projects over the ground-tube.
7. Adjust height of the bricks/block(s) with the shimming material to make a fulcrum, so that the lever *just* squeezes between said fulcrum and the underside of the nail. It should be so close that the lever is nearly vertical.
8. Push down on the lever. The nail, plus rib, will be forced upwards, with remarkable ease. Remember: the longer the lever, the easier the lift. If the nail bends, and you can get it out again, replace it with a stronger one. If you can't get it out again, serve you right. Now you'll have to go all the way back to the shed to get the hacksaw, which you should have brought with you in the first place, as you well know. What's more you now have to be careful not to damage the brand new skin with the saw – and all because you didn't pay enough attention to the thickness

of the nail. A 4in nail *should* be adequate, but if you can get a thicker one through the little holes in the rib, go for it. The thicker the better.

9. So ... after that little drama, and when you're quite ready: maintain the tension on the lever, so the *lower* nail hole is held just above the top of the ground tube, and slide the spare nail through said lower nail hole. You may need an assistant for this, although it's actually quite easy to use your knee or foot to keep the pressure on while poking the nail through with your hand. If your foot slips, you're in trouble of course as, counterforces being what they are, you are quite likely to be slung straight out through the plastic and into the nettles, to the great entertainment of the serried ranks of neighbours, Jersey heifers, stray dogs, etc., who have come to see the fun.
10. Remove the 'old' nail from the uppermost hole, for use as the 'spare' for the next rib.
11. Stand back and shriek with delight; or punch the air; or play the solo from 'We Are the Champions' on your trusty air guitar, as the spirit moves you.
12. Repeat as necessary, getting faster and faster as you go.
13. Feel really *really* smug as the last nail slides home and the tunnel skin booms like a kettledrum when you flick it with your fingernail.
14. Crack open the can of favourite tipple and muse upon Archimedes' potent observation: 'Give me a long enough lever, and I will move the Earth.' Alternatively, ponder upon James the Builder's more profoundly realistic observation: 'Give me a long enough lever, and I'll lose it.'

* * *

If everyone could put up their own 30ft x 14ft polytunnel the world would be a happier place. A family of four could grow just about all its greens and in-season exotica like toms and peppers and salads; and finger-licking fresh, too. This would mean a great improvement in national nutrition and a consequent huge reduction in the National Health bill, plus an even huger reduction in greenhouse emissions as fewer megatons of carrots, peas and cabbages would 'need' to be hauled across hundreds of miles of congested roads from farm to pack-house to distribution centre to supermarket and eventually to a kitchen: possibly your kitchen, which might be just round the corner from the original farm.

You can grow anything in a tunnel. It's just a question of timing and relative value. Spuds grow very well on a field, so we plant just a few earlies 'indoors'. If you are really efficient, you can grow one spud plant in a tower of old tyres or a wheelie bin, topping up with rich soil and compost as the leaves poke through. A big crop from a small area.

You just need to keep an eye on ventilation and watering. Half-netted doors should let the air through and keep the birds and butterflies and cats out, but watering might need more attention. What we do is to dig in as much well-rotted compost as we can, then really drench the whole lot. The compost will retain a lot of moisture. Then we plant and let the surface dry off, to deter slugs, moulds and mildews. To keep the surface dry you might like to consider burying a few lengths of semi-permeable 'seep-hose' which you can just click into your irrigation line. This will release a very slow ooze of water at root level. Brilliant stuff.

More simply, sink a plantpot or half a perforated pop

bottle next to each tom plant, and water into that: efficient delivery of water, plus dry surface.

Restricted watering means sweeter and more flavourful melons and strawberries. It also means a smaller crop, however, so take your pick.

The other delight of a tunnel is that you can grow early and late crops in it, expanding your season by a month or more. You can even buy special lettuces that grow (very … *s…l…o…w…l…y* …) over winter.

One day I might see just how much I can grow in one tunnel over one twelve-month period, immediately replacing one crop with another so that the soil never remains bare. I think the results would be truly amazing.

Overheating can just occasionally be a problem. Some people whitewash their covering to reflect light, and shade the crops. Personally, I think it's a much better idea to grow a grapevine along the ridge pole. The leaves give quite a bit of overhead shade, and you can always train runners down a rib or two if you really need more. If you need less, just prune back. And the leaves can come in handy on Greek Nights at your local Balti house, and, so they tell me, on Real Man camping trips when oak leaves may prove to be just a bit fiddly.

The dug-in compost gives enough food for a season or more of intensive growing. We've tried commercial organic fertilisers (made from dried muck of some sort) and calcified seaweed, but we really don't need it. Perhaps once in a while we'll use a shot of liquid seaweed as a foliar feed, but only rarely. We prefer our own comfrey feed, anyway. More on this below.

To keep the soil 'sweet' (i.e. non-acidic) we chuck an occasional bucketful of ground limestone into the compost at sowing time. It's a slow-release form of lime (slow-release is an essential quality for any organic additive) and it's hard to overdose with it. We had ten tons dumped into the back of the black barn when we first moved in. Unfortunately, the barn has filled up quite a lot since then, so access is now a bit of an adventure, elbowing past defunct washing-machines, old freezers, useful boxes, tubs, bottles, planks, etc. But the lime rock will never go off.

Our tunnels are magnets for rabbits. They kick, bite, punch, batter, and tunnel their way in. The best defence seems to be to scatter a shovelful of what dogs do best. They still get in though, even in the off-season, and dig holes to nest in. Last year I discovered the beginnings of yet another tunnel and collapsed it with my heel, to discourage further efforts, and also to protect my own ankles against a sudden twisting trip. Weeks later I began the spring dig, and found four perfectly preserved little bunnies in the collapsed hole.

Now here's a moral dilemma: I didn't know the kits were in there when I collapsed their nest. However, what would I have done if I had known? Could I have left five (or more) voracious creatures to destroy every plant in the tunnel? And if I did, how long would it be before there were fifty or a hundred of them? Where would our own food supply be then? What would you have done, dear reader?

It's not easy, is it? Especially when your *living* literally depends on it.

* * *

Russian comfrey makes a fine foliar feed. All you need is a forty-gallon oil drum, fifty lusty comfrey plants, a big rock, a foot of narrow plastic tubing, a one-gallon plastic bottle, and a small hole in the ground. Easy.

Place the drum upright, downhill from the comfrey plants (because it's easier to carry stuff downhill than uphill, yes?) and next to the small hole in the ground. Stuff the plastic bottle into the hole. Use a nail punch or your martial arts skills to pierce a small circular hole at the base of the oil drum. Push the plastic tubing *tight* into the hole. Shove the other end into the bottle. The idea is that liquor forming in the drum will drain into the bottle, which is why the bottle is downhill from the oil drum.

Scythe as much comfrey as you can stand, and stuff it into the drum.

Now comes the entertaining bit.

Climb into the barrel and jump up and down on the comfrey.

Anyone with any imagination at all, or who has tried using a child's skateboard when drunk, will immediately see the comic and surgical potential in this. First of all, actually getting into the barrel is not nearly as easy as you might think. And, heavens … you don't really want to have to get a *stepladder*, do you?

Seriously, if you ever want to get a humdrum party going, roll in the plastic water butt from the garden, or better still, a proper steel drum with a nice jaggedy rim where the top has been battered off with a dull chisel, and offer suitably silly prizes* for anyone who can get into it and stand upright like a Jack Tar in a crow's nest, with no assistance. Diving in

* A child's skateboard is one suggestion.

does not count. Clambering half-in and falling over does not count. Neither does tilting it and sort of hopping round the room, bleeding everywhere. Straight in. Standing upright. Then half fill it with cushions and try again. Six-footers are automatically barred unless they have one leg in plaster already.

When stamped and tamped, jump out (again, not quite as easy as it sounds) and add more comfrey. Jump in again, blah-blah.

When the drum is absolutely full and your legs have turned to rubber, place a 'chaser' on top of the comfrey. This will normally be the top of the drum which you have previously chiselled off, generating tinnitus throughout the neighbourhood in the process. But it could be anything that just fits inside the diameter and will press solidly down onto the comfrey.

You now need to get something heavy to sit on the chaser. An old wheel might do, but it's a bit lightweight. We use a huge quartz rock, bound round with baler twine* so we can get it out again when we want to top the barrel up. Even with the twine, it's still something of a pantomime, leaning over into the barrel trying to haul a 50lb rock vertically upwards, bent double, with both feet off the ground.

You might like to consider this variation for later in the evening, when your party has warmed up a bit and the paramedics have left.

---

* When we retire I will name it '*The World in Bondage Number 8 (Hercules)*' and sell it for a fortune to Charles Saatchi. Come to think of it the oil drum should be worth a bob or two as well: '*Life's a Bitch*' sounds pretty good to me: or perhaps '*Oil Drum*', if I'm feeling ironic.

Barrel Tumbling Craze Sweeps Islington
Samaritans can't cope

After a week or so the comfrey rots down and produces a thick black coffin fluid. Sometimes it's thinner. Either way, dilute it 10:1 and use as a foliar feed.

We once made so much of this stuff that I started storing it in an old milk churn. This fitted neatly onto my bike trailer, and I hauled it slowly up and down the field, delivering draught liquid fertiliser to anyone who fancied a drop, via a nickel bath tap screwed into a hole punched in the churn.*

The remaining sludge from the barrels goes into the compost heap.

I guess you *could* scale the whole process down and use a bucket and a teacup, but it wouldn't be half as much fun at parties.

An excellent use for comfrey leaf is to let it wilt, then lay it in layers under your seed spuds. As it rots down it provides a bit of extra warmth, then a fertiliser boost for the crop.

* * *

Speaking of compost, and barrels …

The best starter for a compost heap is urine, of which we waste millions of gallons every day (between us). Why not get yourself a five-gallon plastic drum and stick it next to the loo, preferably with an amusing label on the side of it? My own

---

* IKEA have since bought the idea off me as a wine-dispenser for ya-ya parties and Masonic feasts. Probably retailing at £299, inc. five gallons of fine Muscadet.

label was 'Big Boy', peeled carefully off a new watering can. I'm sure many amusing possibilities will occur to you. 'Vitamin P' is a useful fallback.

Use is simple: Gentlemen: direct jet A into barrel B. Then screw the top down *really* firmly and as quickly as possible, for reasons your wife will be keen to explain, possibly at some length. After a while, pour it onto the compost heap, preferably diluted. And if you are troubled by cats in your garden, and have a high-powered water pistol, you may think of some other use for it.

Sheep shifting

Our livestock kept thriving and multiplying. In October 1991 we took up our bamboo staffs and trekked twenty-three sheep across to Ken's for dipping. None got lost, even without collars, because we were by now pretty wise as to the Ways of

Sheep. The key is that they will always follow a leader, unless spooked. So all you need is a magic bucket with a few 'sweeties' in, trailed a foot in front of the leader. She will follow it; the rest will follow her. Simple.

We'd also learned how to shoo sheep. If one escapes, her prime aim in life is to get back with the others. Thus, all you have to do is to assist her, using the minimum of guidance. This is not done by rugby-tackling her or hauling her by the ear, or zapping the poor lamb with a taser. All you need to do is to take a bamboo in each hand and slowly approach. Don't get too close or she will spook and you've lost her. Pause. Advance. Tap the ground with a bamboo. Perhaps wave it slightly. She will definitely move away from it. In no time at all she'll have gone where you both want her to be. A proper farmer doesn't need the bamboos.

Of course, singling a sheep *out* of the pack is an entirely different matter. If you have a big flock, you'll need one of those wild-eyed Border Collies who'll do it for you in twenty seconds flat, then dash off like a water-pistolled cat to round up the chickens and ducks and anything else he can find (snails will do; or a pile of bricks) before tea-time. If you have only a small flock, the magic bucket is quicker. Just lead them all into a pen and hoik out the one you need. 'Oy ... Brenda ... yes, you ... over here. Now! Oh ... well, never mind ...'

Exciting New Board Game Offer!!

'Can the Hippie Catch the Sheep?'

Just Send In Eight Tokens With The
'Hopeless Hippie' Picture On

– 25 –
Milking

We learned to hand-milk on Daisy. When Ken led her down the ramp of the horsebox he'd brought her in, she was clearly the size of a boss hippo, but after a couple of weeks of getting to know us she shrank, pretty much to the size of a large ginger dog. Curious trick of the light, I thought.

Daisy didn't mind being milked too much. Ken showed us how to place the bucket; exactly where to sit; how to grease the teat; how to gently squeeze, from the top downwards; how to direct the jet of milk; how to gently repeat the process over and over, head pressed to warm flank, ruminatively; then how to remove the filthy hoof from the three inches of milk in the bucket; and how to start all over again.

What he didn't teach was how to keep your sense of humour after your third trip to the sterilising fluid in the kitchen, in the cold, in the rain, in the dark ...

For the fourth time you go through the routine: pat on the neck; cheery banter re the weather, likelihood of Swansea winning the cup, etc.; a quick bump upwards on the udder to simulate a calf butting for its lunch; a gentle stroke, and squeeze and squirt; and squeeze and squirt; and squeeze and …

… begin the long trek back to the kitchen *again*, muttering Teutonically.

Once the old cow did it to me five times in a row. Why, I don't know. I was doing everything right. She had a wad of fragrant hay and a bucket of concentrates. She was warm and dry, and was having the pressure on her udder skilfully reduced, by a friend. Bliss. Then wallop: back foot in the bucket *again.* Perhaps she liked the nice feeling of warm milk round her toe? Or 'cloven hoof' …?

I tried hard to be patient, but five times was just too much after a busy day blocking holes in hedges and humping breezeblocks around. As the hoof clanged in for the fifth time I hurtled upwards off the stool, grabbed the bucket and yelled, 'You stupid bloody cow' at the top of my voice, and hurled the filthy milk all over her.

Please don't tell me I was wrong to do it. I know I was. Poor helpless beast … yes I know. Oh, those big cow eyes, all innocence and hurt, like a Wronged Royal on *Panorama.* 'Yes, you! You are a stupid and horrible old bloody old cow, and I'm sick of you! You can milk yourself from now on. I'm going to get my own hay now. Dinner. Own *dinner.* I'm going, anyway.' And out I flounced, with just one backward glance, bovicide in my eyes. But … she looked so profoundly bewildered and *shocked,* that I had to go straight back in and give her a cuddle. 'Look … sorry ... Didn't mean what I said. Just a bit tired. Why don't we give it one more try? Hmm? C'mon. Give us a kiss. No, not really. Oh stop it! No. Gerroff …'

Amazingly, it went without a hitch. Somehow we'd broken the 'Oh-milking?-must-jam-foot-in-bucket' spell. Cows have very few behaviour patterns, all very slow, and even slower to change.

When it does go smoothly, milking is actually very pleasurable; a sort of gentle communion that goes back thousands of years. We remember how closely and immediately we rely on Nature for our living, and how we are linked to endless generations of peasant farmers.

They say a cow will give up to fifteen per cent more milk if you play a little gentle music while milking, but fifteen per cent *more* was the last thing we wanted (and anyway I never did get the hang of playing a guitar with my teeth). At peak times we had more milk than the six of us could possibly use. Anne made butter and cheese, and we still had a surplus.

But I was never as good at milking as Anne, and she gradually took over. Then she started with a bit of arthritis in the wrists, so we bought a little electric milker. This saved her wrists, but otherwise was a waste of time, because of the lengthy rigmarole of setting up and cleaning and rinsing and sterilising afterwards. If we'd had six milking cows it would have been wonderful. But not just for one.

An added complication was that big flies used to bite Daisy's udders and teats and leave big seeping sores, which meant that the vacuum suction cups of the machine were painful for her. We tried ointment but she licked it off. We tried masking agents like citronella and garlic oil, so the poor beast stank like a downmarket Turkish brothel, but she licked that off as well. Meanwhile, she was in some distress, as the open sores attracted more and more flies.

Anne found the solution: a DIY brassière. We tried a couple of prototypes, using various cushion covers and carrier-bags harnessed with colour-co-ordinated lengths of baler twine. She licked and kicked them all off, even the Harrods one, which we thought looked rather cute. Then Anne thought, 'What do we have that has a bowl-shaped part, and two long parts, preferably stretchy, that can be tied over her back?' Something that will be snugly uplifting, without buckling her udders too unfashionably? Ladies? Yes … a pair of tights. Absolutely perfect. The sheep sniggered a bit, but what a small price to pay for freedom from those poisonous flies.

When Daisy finally had to go, her daughter April took over. We tried hand-milking her, but basically she was the Audrey Hepburn to Daisy's Marilyn Monroe, and I think it hurt her trying to tug on her tiny back teats. She had the milk alright; it was just the mode of access that was a problem. So it had to be the machine again, time-consuming though it was.

Upon fetching the cow for milking

Don't look at me like that, my girl.
I know that look:
Dumb insolence:
slack-lidded, middle-distance
Dumb
Cud-chewing
Insolence.

Oy! Hup!! Come on!!
It's milking time,
as well you know …
Come on!!

A pause in chewing.
GULP.
Another pause.
A tiny widening of the eye
As the next green gob of pulp

Escalates up.
Then chaw chaw chaw ...
The endless macerating meditation,
Grinding slow but sure
As the wheels of time.

Oy!! Come on!!
I know you can hear me ...
even if you're not listening.

And don't fart at me, either.

* * *

April continued to give us plenty of what the land needed: grass and shrubbery in; fertiliser out.

During the summer, this just flops or sprays wherever she happens to be standing, but over winter, when she's locked in her shed, it gradually builds up into a thick warm mat, as long as you help her out ...

First of all you scatter a bale or two of straw over her concrete carpet, then wait for her to christen it, randomly and frequently. The straw soaks up the pee and the poo mixes in with it. It starts composting in situ and the heat given off will keep her suite warm.

The sloppy stuff will probably not be scattered evenly, so every now and then you might like to turf it round a bit to level it off. You add more gleaming straw as and when needed. Gradually the mix of straw and dung and pee builds up into a thick spongy mattress called 'deep litter'. By springtime it might be a foot thick, or even more. That's a lot of muck.

One sun-shafted and song-thrilling morning in May, as modom is finally let out of her quarters, and skips off to do her little vernal dance and arabesques across the greening meadow, you tool up for the prospect of mucking out a ton or so of proto-fertility from the shed.

It's got a music hall reputation, like 'smelly' compost and ogress-mothers-in-law, but actually it's not a bad job, as long as you've used enough straw over winter. It does smell a *bit*, but the straw will have caused most of the slop to compost down. And proper compost does not smell.

Your 'problem' will probably be that the dungy mass has been endlessly stood on by a heavy beast with small feet: it's as if a team of rugby forwards has spent several months in there, wearing stilettos. The mass is … compact.

Llanelli Scarlets RFC Show Record Profits Thanks To 'Nice Little Earner' Says Manager, But Lips Sealed Over Details

Your job is to somehow lever and lift this compact mass from the floor of the shed and into the muck-spreader. However, the front door to April's shed is only slightly wider than yours and mine. You can't just back the muck-spreader in. Everything needs to be forked out of the door, one down-lift-hup!-backswing-and-*fli…i…ing!* at a time.

* * *

We've learned a new respect for the people who designed the farmer's range of hand tools. Each one has evolved for a specific purpose. If you don't believe me, ask for the morning off and try shifting half a ton of compacted cow muck out of your kitchen with an ordinary garden fork. You'll break your neck, I guarantee it, before you've got the first forkful into the spreader. Surprised? Read on ...

Now try it with a pitchfork. It's a bit easier, isn't it? Now why's that? Take a look at the tines ('the prickles', if you must). A garden fork has close tines, quite wide and flat. A pitchfork has only two thin tines, widely separated and rounded. These are wonderfully efficient for lifting light airy wads of hay or straw, but not really strong enough to shift clumpy muck. Better than a garden fork, though.

Common sense says that a tool which combines the strength of the garden fork with the slenderness of the pitchfork must be good for muck. And so it is. The 'muck-fork' has four widely separated pitchfork-like tines, and looks like an anorexic garden fork. It is an absolute joy to use in all three aspects of the job:

1. its slim tines plunge *into* the muck like hypodermics;
2. it has the strength of construction to *lift* a good wodge of stuff; and
3. the tines have the slippery slimness that allows the muck to actually slide *off* the fork when you swing and chuck the load off it.

What is likely to 'break your neck' when using a garden fork is not the weight of the load, but the impact of you crashing into the muck-spreader as you sling-and-chuck the muck off the fork; or rather *try* to sling-and-chuck it. You lift, backswing, aim ... and *fli...i...ng!* it away from you, with impeccable force, direction and style ...

But no, mate. It won't 'fling', not even a bit. Instead, it just drags you after it, still skewered on the fork like a sausage on a stick. So in fact you *fli...i...ng!* yourself across the shed, stumbling across the rutted dung matting, out through the doorway, and straight into the sharp bits. Even if you're not injured, this is the most debilitating and frustrating work known to man. Every chuck takes ten times longer than it should.

A muck-fork is what you need. Accept no other.

* * *

An added complication may arise if modom has been careless with her hay. This is quite likely to happen, as cows are not tidy by nature. She will rootle and tunnel into her dinner, turfing it all over the place. Some, or a lot, will end up underfoot and so become instantly inedible. Thus you have lost some expensive feed, and have also set up a trial for yourself come the spring.

Straw soaks up liquor, and rots beautifully. Hay doesn't. It just gets wet and matted. It is almost impossible to get a fork into it, and even harder to lift off a forkful, because the long stems have interwoven into one impenetrable rat's nest of faecal felt. So instead of a compact mass of dry, solid semi-compost, you are struggling with a stinking, dripping, gigantic wig, vainly trying to haul a few hairs out of it. It will take hours to move it and will reduce you to using several rude words. It will. I've done it, so I know.

The following year, you will find a way of making sure that modom can *not* spread her lunch everywhere like a naughty two-year-old. I've done that, too.

Just by the by … if you run short of straw, *Daily Telegraphs*, ripped into ribbons and scattered gaily in rustic swathes and buntings about modom's boudoir, will make pretty good bedding. It holds more gravy than you'd think.

Occasionally we get hold of a bag or two of shredded office paper, both plain and coloured. We tend to keep it for April's Christmas decorations and waft it everywhere in great cumulus billows. She is delighted, and *eats* most of it. Don't ask. Perhaps GCHQ should employ a herd of Jerseys and save a fortune on shredders.

That's it: the secrets of mucking out.

On the What-To-Wear? front, see Chapter 27 for some thoughts on footwear. And be warned that most of your clothing will need at least a serious airing afterwards, if not a downright washing.

On the Strictly Practical front, you can put raw muck on the land, but it tends to rob it of nitrogen. It's much better to compost it, which means aerating it as well as you can, covering it and leaving it for a year. After that it will be much reduced in volume (good news for your back) *and* of higher nutritional value per unit.

Bikes and herbs

We finally gave up on another potential diversification scheme. When we came down here we brought all our old bikes with us, because surely we'd be able to let them out at a modest fee to all the paying-guest visitors?

We had Dad's old bike – the Raleigh he'd been given for becoming Victor Ludorum at school in 1937; our own flashy Falcon ten-speeds; another sporty job with a neat set of close-ratio racing gears; and one or two others: my ice-cream trike, reduced to single components and spread around several boxes and heaps; an old butcher's bike which I used to pedal Paddy to primary school on, perched on a seat from a scrap Fiat 500 which fitted into the basket-frame as if tailor-made for it; and a couple of old friends: Anne's Sturmey-powered Hercules, and a strange anonymous black thing from the 1930s, built like a rhino and with a dynamo the size of a Guinness bottle. And the Pashley tricycle which Anne used to ride round West Bridgford, with tiny Cait strapped into a cut-down washing-up bowl atop the rear basket.

But no, we never let out a single one. This was partly because all our visitors had nippers with them, whose only dream was To Go To The Seaside: too far to cycle; and also because our drive is just too steep. It put off even us. It also put off every single visitor. Lance Armstrong took one look and turned on his heel, sobbing.

Actually, we did get out once, but it wasn't as much fun as we'd hoped. Local boy racers hurtle round the back lanes at stupid speeds because, obviously, nobody would be stupid enough to be out on bikes, would they?

So the bikes just gradually got forgotten, and gradually rotted. Pity.

* * *

The births of spring '92 brought us up to thirty-one sheep, which is quite a lot for a small place. Anne built up a billowing stock of fleeces in old oats sacks, ready for carding, washing, spinning, dyeing, and knitting into jumpers, coats and yet more flipping cushion covers.

But while the sheep boomed, so to speak, the cows didn't. April never really took to the idea of being a foster mother, which was one reason why we gave up buying in poor little bull calves from next door. Pity. But we had managed to give half a dozen or so a bit of a life.

* * *

By now I'd had the ME for some seven years. It came and went with no discernible pattern to it, except that it seemed to be gradually tapering off in terms of severity. After the first attack I'd been more or less bed-ridden for weeks, then just plain exhausted for several months. After that, I'd been fine. This pattern repeated in the second year.

In the third year it came back earlier and lasted longer, but wasn't quite as severe. And so on.

By 1992 it was with me more or less permanently, but I was only rarely reduced to bed for more than a day or two at a time. I could work, after a fashion, but the constant uncertainty meant we couldn't really plan anything, including a proper planting and cropping regime.

What would get me well again?

We thought we'd try Chinese herbs ...

The nearest qualified dispenser was forty miles away, in Swansea. We took a trip on the train.

Rob welcomed me and sat me down. Then he looked carefully at my eyes and my tongue. Then he took my wrist and felt for the three traditional pulses Chinese doctors deal with. (Why, I wondered, don't Western doctors do this? I still don't know the answer. If you ask them they just look blank.) Then he asked me a few questions, and that was that.

Back in the dispensary he took a big paper bag and stuffed it with twigs and bits of odd rubbish obviously shovelled up from an unsterilised forest floor somewhere. 'Boil this for twenty minutes, then drink a cupful every four hours' (or some such). 'No acupuncture?' 'No ... not appropriate.' 'No focusing on the infinite, and ...' 'Just boil these and drink a ...' 'But ... it's just a bagful of scraps and offcuts.'

He then went through each shrivelled morsel one by one and explained what they all were and what they were good for. For someone who thinks of herbs in terms of a delicate sage tea, or a soothing chamomile ointment, this was something of a shock. But ... if it had served the Chinese, the inventors of paper, gunpowder and noodles, and, quite probably, everything else from football to Pythagoras's theorem, for thousands of years ... well, it was good enough for me, particularly as the Western treatment appeared to be 0 (which the Chinese quite probably invented as well, unless the Indians really did beat them to it).

Back home we boiled the stuff up. God, what a stink. I'm going to *drink* this?

And I'm afraid it tasted even worse than it smelled: a bit like essence of liquorice, and extra-sour Lime'n'Iodine poultice rendered for days with a full rugby kit, stilettos and all. Just *horrible.*

I stuck with it though, right to the bitter dregs and an aftertaste that lasted for hours. Then went back for more. I was definitely not going to miss out even on a placebo effect if one was available.

It was a pricey business, travelling to Swansea each time, although I spread the cost by getting to know all the second-hand bookshops. But it would all be worth it if ...

However and alas, the herbs seemed to do no good. None that I could discern at any rate. Perhaps they stopped me getting worse? Who knows.

Anyway ...

– 26 –

Of WWOOFing and WWOOFers

'WWOOFer' is a term that needs careful handling. Should a smallholder casually mention in the pub that he's 'having a couple of WWOOFers over for the weekend', he must be prepared for the odd raised eyebrow, and possibly a stream of free drinks from that biker in the corner with the Saddam Hussein moustache and the little leather shorts.

Of course 'WWOOF', as every organic farmer knows, stands for '*World Wide Opportunities on Organic Farms*', and WWOOFing is a splendid thing to do. It gets townies out into the countryside, where they can get a bit of peace and quiet, and learn something of country skills and organic theory and practice; and hard-pressed farmers can get a bit of help for a couple of days with long or difficult jobs. Everybody's happy, especially as no money changes hands.

The system is a model of non-bureaucratic efficiency, from which many a manager of systems and people could learn.

1. A farmer registers as a WWOOF host.
2. A punter becomes a WWOOF member by paying an annual £15 subscription and receives the Host List.
3. The WWOOFer contacts a farmer, to check if s/he needs any help.
4. The WWOOFer arrives on the arranged date, and is fed,

watered and sheltered (I almost wrote 'bedded' but thought better of it) for the agreed time in exchange for work around the farm.

5. The WWOOFer leaves. Both s/he and the farmer are expected to report back to WWOOF HQ if dissatisfied in any way. Dissatisfaction is rare, but if registered, then stern phone calls may be expected, and possible expulsion, with bell, book and rechargeable torch in extreme cases.

It's a brilliantly simple and effective system. Over a five-year period we had a stream of likeable and helpful WWOOFers descend on us in ones, twos, threes and fours. There were only two duds out of the lot. One was a whinger who got on everybody's nerves, including the other WWOOFer's; and the other one was exploiting the system by 'Bed & Breakfasting': eating the maximum of good food in return for the minimum of bad work. We checked around with other local hosts who apparently had all had similar experiences with him. One phone call, and he was banned. Simple.

He also had the alarming habit of turning up in the kitchen at half past six in the morning, to do his 'dancing'… in his underpants and T-shirt, and wanted Anne to join in. Er … no.

Just a harmless eccentric, you might be thinking, but it didn't feel like it. There was a slightly sinister feel to him.

His parting act was to walk off with my belt, which he denied stealing, as he unbuckled it and handed it to me after I'd chased two miles up the road after him.

Our only other *slightly* negative surprise came from a very likeable family of four who were shortly off to Israel to join a moshav, and were sensibly using the WWOOF experience to get used to a different way of living. They literally didn't know what a tomato plant looked like when they arrived. We got on fine, and they worked hard. But when they left, their caravan was like a bomb-site. Communal living? Steep learning-curve?

Everyone else has been a complete delight. Mainly twenty- and thirty-somethings out from the city, they bring news of new trends and developments (which always seem to begin in cities, don't they?) and often have enthralling travellers' tales to tell. Where else would I be able to listen for hours to stories of walking through the Philippines? Or talk to a draft dodger from pre-Mandela RSA?

We had WWOOFers from all over Europe, not to mention NZ, the USA and Oz, and they all put plenty of happy work in. One charming fraulein smoked a pipe and made her own nettle soup, and refused to stop for coffee breaks because it was 'not efficient'. No kidding. But she changed her mind when she realised we worked an eight-to-eight day. She spent her last 10p sterling calling us from Cardiff airport to say thank you for having her.

Choosing the work for WWOOFers is a bit of an art. They all want to do something interesting like ploughing or skinning rabbits, or learn a new skill like hedge-laying; but the farmer desperately wants half a mile of spring onions weeding. So we both have to compromise: a bit of this, a bit of that. With two helpers, you can get a surprising number of jobs done over a weekend: washing the polytunnel skins, pruning the orchard, laying a new length of hose, *and* mucking the tunnels. (Oddly, you never run out of jobs to do. A curiosity of farm life, I always think.)

We eventually had to give up hosting when illness caused us to cut back production, and the cost of employers' insurance skyrocketed. I do miss having new people to meet and learn from.

Two WWOOFers live on simply by virtue of their names …

We already knew a 'Farmer Giles' and a 'Black Smith' (a farrier from Jamaica). WWOOFing then brought us two more 'unlikely' names, held respectively by a policewoman, and a girl from Alsace. We thus have met an Officer Dibble and, wait for it … an Alsatian WWOOFer.

The Alsatian worked doggedly but kept chasing the sheep … no, seriously: she was an excellent guest and had a local surname, 'Xeuxet', that allowed me to make my one and only joke in French: 'rien ne succède comme Xeuxet'. Time to go, I think.

s.a.e. to WWOOF, PO Box 2675, Lewes, East Sussex BN7 1RB. Or try www.wwoof.org.uk (email: hello@wwoof.org)

* * *

A couple of WWOOFers erected a length of fencing for us. They'd been on a proper Country Skills course, which is more than we ever had, and did a marvellous job of running a twenty-metre strip down a steep and variable slope. This is tricky stuff, believe me, if you don't want ugly pleats and saggy bits in your netting, leading to a rash of silent nudges and smirks from the sheep. This stretch is by far the best bit of fencing on the place. I wish I could remember their names to thank them here.

Even Townies Can Put Fences Up Shock
Courgette World Rocked to its Marrow

Some foreign visitors had tricky accents. But that was OK as I used to teach English as a Foreign Language, and enjoyed the trip down memory lane. I was even able to re-educate a few of them into saying 'can' and 'carnt', instead of the irritating and baffling American version of 'can' and 'can' which they had been taught at school.

Oddly, the WWOOFer who tripped us up most often was a pleasant Kiwi lad. He was young enough to identify more with our near-teen Cait than with us oldies. One evening he was bored with watching *The Bill* and asked Cait if she'd 'like a game of scribble'. Cait looked puzzled, as did we. We could sense that she was feeling slightly talked down to. Scribble? At her age? Colouring in?

'Scribble?'

'Yeah ... dontcha hev it over here? A little board with letters on tiles ...?'

But the WWOOFer who sticks most clearly in our memory was the utterly remarkable Sara, from Amsterdam. At first glance she was tall, slim, blonde, attractive, twenty-something, smiling, and confident. That was surely enough?

But no ... she spoke five languages (including Finnish, of all things) and was currently doing a postgraduate course in international relations for her job with a UN refugee project. Pretty impressive.

But there's more ...

We offered her a choice of work and she picked the heaviest: digging out and barrowing away a ton or so of earth, in cramped circumstances, from up against the utility room wall.* I offered her a break, or a spell at something lighter, but she wouldn't hear of it. This was her task and she was going to finish it. And so she did.

What do you make of that? Wonderful stuff.

There's more.

She put her tools away, smiled, and said, 'Now I would like to make supper for you.' And so she did. A fragrant cauliflower curry.

On the table we found a bottle of white wine, decorated with a couple of little streamers, handcrafted from bits of coloured scrap paper, and an improvised card saying 'Happy Birthday Anne'. We must have mentioned the birthday at some point during the day … or did she just notice the cards on display? I can't remember.

Wasn't that just remarkable? What a guest!

To finish the day off, she played us a selection of Bob Marley songs, sung sweetly to her guitar. No, she wasn't juggling three ferrets with her other hand, but you can't have everything, can you?

After the endless blather in the media about micro-talented 'mega-stars' and self-regarding politicians and pointless 'celebrities', isn't it deeply refreshing to realise that once in a while Life sends you a genuinely superior being, to bring a smile to your heart and refresh your faith in what Man might be?

* * *

* On the *outside*, I hasten to add.

WWOOFing is a great way of getting round the country, meeting new people and doing a bit of meaningful healthy work. And you can now do it in fifty-four other countries, too, all the way from Finland to Hawaii, and back via Uganda, Slovenia, Mexico, and Nepal, if you have a poor sense of direction.

* * *

Perhaps we could have drafted in a stream of willing and talented WWOOFers to help with one of our Big Projects: to find a way of getting spring/field water from the sump just above our top field, *over* the slight intervening rise, and to a reservoir system at the top of the field itself.

That slight rise was the problem. No doubt we could have done something clever with siphons if the height differential had been great enough, but it wasn't. My guess is that the original height and the ending height were almost identical. Thus siphoning wouldn't work.

So what would work?

A windmill!

I drew up rough plans for a conventional fan blade, and a less conventional scoop-blade made from a chopped down five-gallon drum, and decided that either design would generate enough power to force water over the hump and into one of a short row of old central heating oil tanks, lined up on the headland; possibly on low plinths, to give the water enough head for watering duties.

Probably the simplest way would be for the windmill to charge a battery which would power an electric pump. Easy … even I could rig this one up. All I would need was a bit of help to get the blade up into the airstream. And there was the problem, as any half-competent mechanic will already have realised …

It has been a failing with me for decades, to not appreciate the simple truth that the clever-techie stuff is rarely the weak point in any sort of mechanical device. It's the *structure*, the ironwork, the frame, the box … which rots and rusts and splits, or just rattles the whole thing to pieces.*

Try as I might, I could think of no sensible way of erecting a windmill over the sump. The land is boggy, sloping, next to a ditch, a hedge, and a fence; and cheek by jowl to the drive. Hopeless, really.

Pity. I really fancied making a working windmill.

A friend helpfully suggested running a power line up to the sump from the house, but that wouldn't have been the same, would it? And someone/thing would be bound to mow/dig/bite through it anyway. *I* would. I just know I would.

Years later it dawned on me that much the easiest method would be to run a polythene pipe directly from the water source that fed the sump … straight down the hill, through the hedge and into the tanks, avoiding the sump completely. Heigh ho. But by then I'd given up on the whole scheme.

And it's far too late to tell me now that the windmill didn't actually need to be anywhere *near* the sump, if all it was going to do was charge a battery …

* Which goes first on the family car, the engine or the bodywork?

The Birmingham job

Alan from HDRA rang me one day to say he'd been asked by a firm of architects to draw up some figures for a job they were doing. Alan didn't have the time, but thought we might be interested? Might be worth a few quid?

Tell us more, brother…

Some very run-down tower blocks in a Birmingham suburb were due for demolition and the Powers were interested in making the most resident-friendly use possible of the land. Obviously new dwellings would be going up on it, but what of the rest of the space? They wanted to know how many families a one-acre plot could provide a living for. Once we'd stopped laughing we thought we ought to do our best for these forward-thinking council folk. Full marks for intention, but not for realism. One acre? A living? *Families?*

How out of touch we have become. The traditional basic requirement for a peasant farmer in medieval Britain was 'five acres and a cow'. That would keep a family in greater or lesser comfort, depending on the nature of the soil. An acre of good soil might have five tons of creative bacteria in it, all helping the plants to grow. They would be sustained by the dung and compost the cow (and possibly the farmer's family) made. One acre of prime Irish land, before the awful famines, could produce up to nine tons of potatoes: that's twenty-five kilos a day, plenty to feed the biggest family. (But what unimaginable monotony … 'What's for dinner, Mum?')

However, the post-industrial 'soil' of Brummagem was a quite different matter. Any self-respecting creative bacteria would long since have packed up and moved out, and I doubt if you could coax half a dozen *bags*, never mind tons, out of an acre of slag and rubble. It would need careful building up before it could become useful again, and that would be time-consuming and expensive. And who would be running this acre? Townies with no experience or knowledge, presumably.

Sorry, councillors … the answer to your question is 'none: zero'.

However … a standard allotment of a sixteenth of an acre, restored to fertile health and then well run and maintained, could keep a family in seasonal veg. Perhaps that would be the way to go, rather than trying to force 'a living' for someone?

Anne and I spent three or four evenings looking into the ins and outs of various veg, and ways of cropping intensively and sustainably. Obviously, 'sustainably' meant that any system must be organic, which meant building in provision for composting (which might actually be quite easy … plenty of organic waste from local households, markets, parks, streets …) Then we considered how to go about educating the would-be allotmenteers.

We drew up a pretty professional report, pointing out as many hidden snags as we could think of (Water supply? Light? Vandalism? Thieving? Drainage? Vermin? Dumping?) and sent it off to the architects. They incorporated it into their own feasibility study, and that was that.

Weeks later I received a phone call. The report had been well received, and what was my fee? We had no idea … let's see ... three or four evenings fiddling about; a day of writing and typing … How about £50?

There was a palpable silence from the other end of the phone. Oh dear ... £50 was too greedy …

'Erm ... the er ... normal fee for this sort of thing is £500.'

What??? 'Oh ... well, that'll be fine. Thanks.'

So that's how the other half live ... they think in hundreds where we think in tens! That five hundred came in very handy indeed. A gift from the gods.

* * *

One thing it came in helpful for was paying for a few more experiments in treatment for the ME. The Chinese herbs hadn't helped, even though they were revolting enough to shock a whale out of a sneezing fit. What else was on offer?

We'd already tried a bit of homeopathy via a trainee therapist. After a couple of treatments she'd died, which was disconcerting.

Then I went to a healer. This meant driving twenty miles each way for her to lay hands on my head while I lay on a trolley in a room that smelled of polish. No good, and pricey in petrol.

Next was another acquaintance who tried a few radionics on me, after a quick dowse with a pendulum, all topped up with a dozen homeopathic rescue pills. No good.

The only alternative system that seemed to help was two visits to a shiatsu expert. Both Anne and Cait thought I was the better for them. Unfortunately the treatment wasn't cheap, and anyway, the therapist moved out of the district after my second visit. (Nothing personal, I believe.)

One way and another, it looked as though this blooming ME was going to be with me until one of us got bored.

* * *

I realised one day that we actually had our 'fair share' of land. There are approximately fifty million acres of farming land in Britain. That's roughly one acre per person. Our family unit of four adults and two kids was farming on about five acres. An intriguing coincidence.

– 27 –
On Footwear

So … you've chosen your headgear: probably a steel-plated woolly hat with a long stiff peak, furry earflaps, and a lobster-tail guttering system at the rear, or something similar (see Chapter 10); and settled on some sort of summer-resistant coat and waxed trousers (see Chapter 14); all you need now is some footwear, unless you're an ex-para or New Age Provisional who regards shoes as fit only for slack-wristed hair-dressers and professional sissies.

Precisely what *sort* of footwear do you need? Well, it depends … Do you intend spending a fragrant hour mucking out the cowshed, or are you weeding kohl-rabi on the field?

For mucking out, nothing can compete with the humble wellie. Gold-thread Blahniks and old flip-flops simply let in too much … let's say 'fluid'. But which wellies? Personally I have three sets, each suitable for different conditions. Set 1 are good for dry rough work: they have lesions in the soles, but are otherwise fine. Set 2 are for working in dew or damp: the soles are fine, but they have assorted splits and snags and teeth marks elsewhere that would let serious water in. Set 3 are all-weather jobs, suitable for anything the Atlantic can throw at us, provided it's less than calf-deep.

Please note that I specify 'sets' rather than 'pairs'. While Set 3 are both black, Sets 1 and 2 are of the harlequin persuasion. One black and one green, to be precise. This is because both members of a pair of wellies never give out at the same time.* Normally, you get a flat in one, but the other one is still good for five thousand miles. So what's the point in junking them both? What's more, if the one with the puncture still works fine in dry conditions, why junk that one either? This is my Green philosophy and I'm sticking to it. I waste less money on new wellies than anyone I know. And one fewer pair of wellies means two fewer blasts of greenhouse gas released in their making.

I don't actually enjoy wearing them, though. I'd much rather wear leather boots if I could rely on them being watertight, which mine never quite seem to be. And boots are a real pain to put on, with all that intricate crochet-work with the laces, and then they're a nightmare to take off after a busy day when you're too tired to fall over, never mind bend carefully forward to unpick with your tattered fingernails the double bow that has somehow cemented itself into a decorative mud-baked swag. Then, when you've teased the knots open with a fork, you can't get the boot off because your foot has swollen and your socks are soaked with … fluid; and foot, sock, and boot are shrink-bonded into one, as the argyle pattern on your instep later testifies. Ugh. Wellies win on this score, every time. Plus you can kick wellies off if you're suddenly caught short in mid-ditch with an attack of the Revenge of the Delhicatessen and Curry House. Wearing boots, you resign yourself to leaving a trail of mud and rubble all the way to the Temple of Release. The alternative is unthinkable.

* Unless you've been trying your luck at trampolining on a bed of nails, in which case you deserve everything you get.

There is a compromise garment for summer wear. Take an old wellie and carefully snip off the leg part just above the ankle. You will now have (a) an attractive rubber slipper, suitable for light and more-or-less dry work, and (b) a foot-long tube of rubberised canvas which snips up nicely to make hinges for shed doors, and patches for leaks in roofs of chicken-hutch, car, etc. If one is fashion-conscious, one might slit open a couple of dozen and rivet them together to form an eye-catching clinker-built poncho.

These slippers work very well. One can even buy them ready made, but to my mind this is rather like buying pre-faded jeans or spray-on 'Kuntri Krappo!' to add credibility to your 'Bloodstained Mangler' 4x4.

The main advantage of the slipper is the freer circulation of air round the ankles. To increase airflow I once tried cutting a hole in a pair just above the heel. It worked well, but the prototype lacked lateral stability and I kept sliding off sideways and twisting my ankle. Also Anne took to calling them my 'sling-backs' which I found unnecessarily aggressive.

I've tried Dutch clogs, too. They are unbelievably warm, even in ice and snow, but you have to learn to walk by rolling forward onto your toes rather than trying to flex the shoe, which can only end in broken bones and serious splinter trouble. I gave them up because they hurt my high instep after an hour or so, but I'm sure the humble clog would be ideal for many people with more normal feet, especially if you glue a decent crêpe sole onto them, to convert them into genuine Dutch brothel-creepers.

New 'Springij-Klogs' Favoured
By Rotterdam Vice Squad
'I Didn't Hear Them Coming!' Mr V Van Gogh

When you come indoors, you just kick them off, fairly carefully, with a loud 'NO!' aimed at the dog.

I also have a pair of huge black Canadian snow boots which get an outing once every year or so. They reach to mid-calf, with heavily ribbed soles. You shove your whole foot in, shoe and all, and fasten up the strange metal lever closures. Snow-chains for the foot. Amazingly warm.

A pair of old slippers are good for light fieldwork. Slip on; kick off. Great. Unlike my 1920s cricket boots, which had been heavily basted for decades with

Meltonian
'Antarctic Blizzard'
Distemper for Whitening Cricketing Boots.

I thought they might make a decent pair of stepping-out shoes, so I tried shining them up with several layers of black polish. Unfortunately, they came out an alarming shade of transparent purple that frightened the chickens, so they had to go. Pity.

My ideal 'shoe for all seasons' would be a pair of thigh-high hob-nailed lace-up leather diving-boots.

* * *

Gosh, don't we take wonderful things like waterproof shoes for granted? I remember watching a television programme once about a group of harmless lunatics who spent a year living an Iron Age life in a big straw hut. What bits of modern civilisation did they miss most?* Yes – wellies.

Incidentally, it's my guess that any community smart enough to smelt iron out of rock was plenty smart enough to also invent the wooden kick-off patten, to raise their leather plimsolls above the mire. These things would not survive archaeologically, of course.

Surprisingly, not one WWOOFer ever arrived with wellies. Usually they turned up in Great Big Boots or … sandals. Curious.

* * *

We noticed one summer that Cait's shoes had suddenly worn alarmingly thin, especially at the heel. Odd. A new pair wore even thinner even quicker.

'What are you doing to your shoes, Caity?'

'Oh, just coming down the hill.'

'Coming down the hill?' *(A picture of her hopping and skipping, lamb-like, down the gently sloping carpet of meadow flowers swiftly gave way to flashes of her scree-jumping at break-neck speed through the broken glass and septic sofa springs of the cwm, shredding her soles en route before crashing into the freezing boulders in the stream.)*

'Well, down the drive ...'

Ahh … 'the *drive*' … cogs began to click into place. Over the previous few weeks of the summer, while we were working on the field, we'd heard jolly shrieks and yelps coming from the direction of the driveway, where Cait was playing with

* Apart from Mars bars?

Edward. We'd paid no attention at the time, but now … come to think of it … those shrieks and yelps had shifted remarkably quickly from the top of the drive towards the bottom …

Aha! … They'd got a taste for speed after the previous winter's venture into hedge-battering, hadn't they?

Cait explained …

They'd started out with an old tricycle, then moved up to an ancient and brakeless tiny bike. They took turns to drag the bike up to the top of the steepest part of our curving and loosely surfaced drive, and then just launched off …

Eighty yards later, at the bottom of the drive, came the tarmac and concrete yard, still sloping. Then came either the dentable back doors of the old ambulance or the far less dentable stone wall of April's shed, or, in extremis, the corrugated-tin wall of the carport which could absorb quite a lot of kinetic energy before bursting apart and allowing the child of the moment to hurtle into the brambles and tree trunks of the cwm, flattened forehead and all.

But even a brakeless bike wasn't fast enough, so Cait knocked up a four-wheeled buggy from the chassis of an old-fashioned springy pram, with a child's car seat creatively attached so that what used to be the seat became the back and the back became the seat. Thus the would-be suicide strapped herself or himself onto the seat, lying more or less supine with the feet sticking out in all available directions.

'Cait … this is lethal.'

'Oh no, we always wear a crash helmet.'

And so they did. She gave us a demo run, toned down rather, I suspect, so as not to alarm parents. I mean, she barely left the ground as she bounced across the diagonal sleeping policeman at the nook of the yard, both heels dragging in the gravel.

'We'll think about it.'

The more we thought, the more it became clear to us that what appeared to be a Cresta Run for Fully Paid-up Idiots was in fact a well thought-out little adventure-sport for ten-year-olds. They could choose how far up the slope they started from, and thus their potential speed; they could slow down and even stop, using just their shoes. The fact that April's wall was not in the least bit dented was proof enough of this. The construction of the trolley was ingenious, and after weeks of use, no one had actually been hurt or even killed. In short: brilliant stuff. 'But you'll have to wear clapped-out wellies. You can use some of my Set 3s if you like. We just can't afford any more shoes. Deal?' Deal.

Fortune favours the brave, as well as the foolish.

Another favourite pastime for Cait and Edward, again unknown to us at the time, was to make elaborate 3-D mazes among the eight-high stack of hay bales. These tunnels were *just* wide enough to crawl through, without any thought given to anything so sissy as airshafts or anything. One of them even had a 'sliding door' to seal them in good and proper.

Who'd be a parent?

Hansel and Gretel in the Hayrick of Death

Two Small Mummified Corpses

'We Were Like Wondering Where They'd Like Got To,' Mumbles Pathetic Hippie.

Souvenir Resin Replicas, Just Three Tiny Inches Long, See pp. 7–8

* * *

Yes, nothing quite beats the feeling of Pure Release when you kick off your wellies on the patio after a busy day, preferably with a cup of tea or a pint of homebrew. Feet splay; toes flex; the air around your socks begins to shimmer and crackle; the dog rams his nose into your wellie, and his eyes glaze over …

Bliss.

Science and life

On my off days I was now reading some pretty thoughtful stuff. Novels held no appeal any more.

What had become much more interesting to me was the mystery of why scientists didn't investigate dowsing and biodynamics and ESP and ghosts and so forth. What was their problem? There are whole libraries full of evidence for all kinds of paranormal phenomena, but they are all routinely ignored or rubbished by men of science. But *why?*

So I read a lot of pop science books. The more I read, the more puzzled I became, because I could find no clear statement of what their problem is. In fact, it took a long time to weasel out the answer, and then 'puzzled' gave way to 'shocked', because I discovered that science ain't what it appears to be.

It is meant to be a dispassionate *system*, which collects evidence and evaluates it without prejudice. But, in fact, science has developed an *agenda*: to promote an unproved (and irrational) philosophy which states that everything in the universe is *only* Matter-Energy. (SfaS).

In a word, science has become dogmatic. *This* is why it

can't cope with ESP and ghosts and so forth: to admit the existence of just one spook would fatally challenge its Materialist dogma, just as Jupiter having moons challenged the medieval Church's geocentric dogma five hundred years ago. Because dogmas must always be defended at all costs (that's what they're *for*), Science resolutely ignores all evidence for the paranormal, just as the Church tried to do with the embarrassment of Jupiter having moons. That's it. Now I understood.

Yes, of course not *all* scientists hold this dogmatic view. But most do. Ask any science teacher.

This shocking discovery made me wonder about the other things science tends to rubbish, religion being the obvious one. As I thought it over, more puzzlements arose. For example, every culture I've ever heard of, at all points in history, has had some sort of belief system that might be called religious, i.e. a recognition that something with a design, order or pattern (like the Universe, for example) must somehow, sometime, have had a cause of this design. Surely this universal belief/understanding means *some*thing? And it's not good enough for us to come over all smug about how primitive all these people were, because these are the guys who invented the wheel, and steel, and maths, and writing. They were no fools.

Something else that struck me was the double standard science applies to its own luminaries and illuminati. When Newton discovered the laws of light and motion he was hailed, quite rightly, as an astonishing hero. But when he turned his attention to alchemy, he was written off as a crank. From genius to loony, just like that. Why? Because he dared to challenge the dogma.

Alfred Russel Wallace, the co-discoverer of the principle of evolution: another genius … until he showed a keen experimental interest in spiritualism. Then he became another loony. William Crookes, Sir Oliver Lodge … men of scientific substance, until they dared to investigate along similar lines. Great original men denounced by small dogmatic men. 'Twas ever so. Every dictator knows that he depends upon the Small Man to keep him in power. I find it saddening that these are the men who seem to dominate the august institution of science.*

I kept on reading and thinking ...

* * *

What has all this to do with smallholding? you may ask. Well, very little on the surface. But my life, as I discovered back in Nottingham before we moved to Wales, is 'a journey', corny though that may sound. Once my life was teaching, then it moved on to smallholding, and now it seemed to be somehow moving on again, even though we were still earning our living from the land. Suddenly I was taking an interest in things that had previously been only of curiosity value, or of no interest at all. Even 'religion' …?

---

* And any scientist who is open-minded enough to think that telepathy, say, *is* worth a proper scientific study is likely to find that he will get no funding and/or be regarded as 'unsound' and not get promoted. Either way, he will have little prospect of having his papers published. When Rupert Sheldrake, a professor of biology, published his book *A New Science of Life*, which dared to suggest that perhaps not *everything* in the universe is just simple material 'stuff and energy', the editor of *Nature* called it 'the best candidate for burning there has been for many years'. The only other people I know of who were fond of burning books were the Inquisition and the Nazis. The thought police are alive and well in the world of science, alas.

Religion? Pah! I'd served nine years of school chapel, and suffered endless hours (500+ in total) of mumbled and meaningless ritual, and boring and meaningless sermons given by chaplains who both seemed just a little too keen on assaulting pupils, one way or another.

Science was the way! And all my science teachers, Materialists to a man, put me on the right road. I even considered becoming a nuclear engineer.

However, thirty years on, I was no longer quite so sure about all this … what was science doing with a *dogma*?

* * *

We were both gradually changing as people. We'd shaken off any 'city needs', like theatres and shopping malls, and now looked upon them as distractions, in the sense that life isn't about getting and owning and throwing away, or catching the latest identikit movie. Life, we had now come to realise, is about growing and developing, and becoming more aware. The miracle of a birth, and of a seed burgeoning into a huge plant: these things made us really think in a way that movies and frock shops never could.

Anne had taken up yoga, and got me into it as well. It wasn't really me, I thought, but I went along with it. Silly foreign nonsense … but it brought out my competitive instinct. *Surely* I could do it faster than Anne? Yep. Every time.

But then I read a book by a yogi and my ideas began to change. I read another of his books, and another …*

---

* If you're interested, these books are by Yogi Ramacharaka: *Fourteen Lessons in Yogi Philosophy, An Advanced Course in ditto*, etc. Still in print after a hundred years.

Soon my old problems with school religion and my new problem with science began to fall into a new perspective.

Anne agreed that these new-to-me ideas made sense, which was helpful. If She Who Understands Things agreed, then I must be on the right track.

Keep reading, brother; keep thinking …

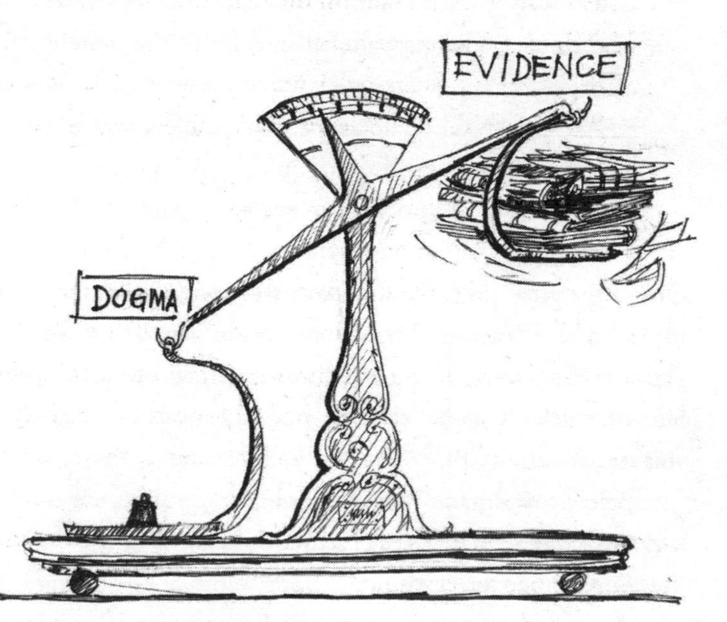

– 28 –
Of Cockerels and Death

CAUTION: This column is unsuitable for those who think cockerels commit suicide for the benefit of chefs.
REASSURANCE: The unspeakable subject of d*ath is over and dealt with quite quickly. But please fast forward to ⏩ if you wish to evade it entirely.*

One of the great joys of smallholding is having room for poultry to potter about in. The ducks tend to waddle off on slug patrol in line astern, like a row of little Captain Mainwarings, but the chickens prefer to stick around home base. Crumbs, you see. Possibility of.

Before foxes made keeping poultry impossible, we usually had five or six hens at a time, scratching up the flowerbeds, at the constant beckon of their big flashy pimp of a Sporting Life cock. He was their protector, largely from other cocks, it must be admitted, and their provider. If he found a tasty snack, like a mouldy slice of bread or an ants' nest, he would call them across. All very domestic and, thus far, nice.

But what about all those other cocks? If you hatch a dozen eggs, you are going to get half a dozen males, and only one of them will get to, shall we say, nurture the hens. The

---

* Good luck.

others are pecked and bullied by the boss cock, like surplus male lions or chimps or, for all I know, bats and protozoa. They skirt around the ladies, brazenly strutting their stuff, but before they can get their chat-up line in ('What's a nice chick like you doing in …' etc.) Big Daddy arrives and boots them into touch.

So then they get, yes, frustrated, and they gradually turn from cute balls of gangly squills into gaudy street fighters. They form cliques and gangs and attack each other on sight. Real Sharks and Jets stuff: skin-tight drumsticks, slicked-back wattles, the lot.

It's very hard penning them overnight in one hutch. It's just one long niggle, and in the morning, bleeding combs and stir-crazy eyes.

Every time, they get like this. And every time it seems kinder, let alone more peaceful and self-sufficient, to cull them.

Let's be clear: 'cull' means 'kill'. I'm not one of Nature's naturals here and find killing anything difficult. But I have grown inured to cudgelling aggressive cockerels. Yes ... 'cudgelling'. I can't bring myself to pulling their scrawny necks. I tried once, and neither of us enjoyed the experience. So I switched to using my old hockey stick, luring the most aggressive members of the posse with grain. Careful aim at the head. Then *whack*. Dead cock. No fear. No pain. But I never looked forward to it.

Once we had a two-tone twosome who were making everyone else's life a misery and would therefore have to go. A visiting friend dragged me down to the pub for a stiffener. We were as brave as infantry when we came back, and tooled up with a hockey stick each. We found Genghis and Kublai easily enough, scattered a few oats, and ... *whack* byebye Genghis. Kublai made it to the concrete path, but we surrounded him, and I delivered him such a wallop to the back of the neck I thought I'd knocked his head clean off. Somehow he staggered into the brambles, obviously on headless-chicken auto-pilot, dead on his feet, and I thought we'd leave him till morning. Any fox that found him in the meantime was welcome.

Next morning Anne said, 'I thought you said you'd got that black and white cock?'

'Yes; so I did.'

'Well who's that then?'

And there he was, struttin his funky stuff; rather slowly and slightly off-beam, perhaps; but struttin nonetheless. I was amazed. He was, of course, reprieved, and interesting to say, from then on he lived a blameless life.

One year we had eight of these miniature velociraptors to dispose of. They'd been pecking each other to shreds and we were all sick of it. The first few went OK, but then they wised up and kept out of sight. I was eventually reduced to lying down in the orchard and picking them off one at a time from their perches through the tiny door of their ark with Dad's big airgun.

Fortunately I'm a good shot, although I did once miss a poor blind waif of a chicken from a range of four inches, but that was an exception. I really didn't want to do it, and my nerves affected my trigger finger, as anyone serving for the match will know about.

I never got used to killing the cocks, even when they were savage little yobs. After all, they were only obeying their natures, and we could live quite happily without chicken crumble. It was never personal.

Except for once ...

Our own hens were mainly Rhode Islands and Dorsets randomly crossed with whatever other odd biddies life sent our way. My parents had a trio of splendid Holland Blues. The cockerel was the size of a small pony, called Dreadnought, and was as sweet as pie.

He had a son, however ...

Unless I'd seen it myself I would never have believed that anything so small could be so supernaturally vicious. From hatching, he was aggressive. His mother loved him, his father doted on him, my parents coddled him ... and he rapidly developed into a rabid little fury. No other poultry, hen or duck, was safe from this cockatrice, even as an infant, and we had to keep Cait well out of his way. He even attacked adults. By the summer, he'd grown considerably and was becoming a serious menace. One day he went for a visitor's child and we all agreed he'd have to go.

I chased that brute for an hour and a half. He knew his time was up but belly-kicked and screeched till the end. With more persistence than I knew I possessed, I finally cornered him and sent him back to whatever genetic nightmare he sprang from. RIP, Ahriman ... please.*

* * *

▶▶ Oh, he was a scamp, that Ahriman, wasn't he? Easy to joke about now, but not so funny when he was likely to ambush and blind someone's child. We've never had another creature

* A friend's multicoloured cock, Joseph by name, also met a short sharp death when the family bull terrier finally got tired of being mounted without so much as a by your leave, let alone a slice of mouldy bread.

remotely as nasty as him. He'd have made us a fortune in the pits of Bangkok.

There was another cock who was also a bit different. He used to pile in with the hens at feeding time, then flap up onto the roosts with them and spend his night balanced on the pole. What made him a bit different was that he was a drake. Very odd. How on earth did he do it? Hens have talons; ducks have webs.

All the other drakes waddled off with the ducks for the night. I wonder what they made of him? Probably something involving the Quackish version of 'pooftah', at a guess.

* * *

One of the joys of keeping poultry is playing Hunt the Eggs. You can build the girls as many nice cosy nest boxes as you like, but there will always be the occasional rugged individual who will wander off to find a home of her own. Sometimes this results in her reappearing one sun-striped morning followed by a train of half a dozen balls of yellow fluff, cute as you like.

Sometimes you never see her again, because the fox has found her first.

Occasionally, she turns up but the fluffballs don't. This may mean that a weasel or mink or seagull or cat or hedgehog or goblin or *something* has dined off her eggs before you can. Or it might mean that for some reason the eggs have died and she's abandoned them; or that the silly hen has just wandered off to lunch somewhere and forgotten her way back. The net result is that once in a while you come across a bit of a dent in some grass behind a slate or under a barrow, with a clutch of eggs in it.

You are strongly advised not to try boiling one up to have with your whole-wheat soldiers. I'm sure you can guess why.

* * *

Birds brighten our days with colour and song. We let our hedges grow up and wild so they have plenty of nesting places. We're not very good on names, but John the WWOOFer is, and he once made a list for us of birds seen on the farm: grey heron, buzzard, lapwing, woodcock, woodpigeon, swift, swallow, house martin, grey wagtail, pied wagtail, wren, dunnock, robin, blackbird, song thrush, mistle thrush, blackcap, chiffchaff, willow warbler, goldcrest, spotted flycatcher, long-tailed tit, marsh tit, coal tit, blue tit, great tit, magpie, jackdaw, rook, crow, house sparrow, chaffinch, greenfinch, linnet, bullfinch, fieldfare, raven, nuthatch, jay, grasshopper warbler, cormorant, and pied flycatcher. Of these, twenty-two were either definitely or probably nesting.

Within a quarter of a mile you could add quail, snipe, curlew, stock dove, cuckoo, skylark, whitethroat, and yellowhammer. In all, some thirty-two species were probably breeding on or around our patch.

The list dates from some years ago. Since then, the numbers seem to have declined. But why? This isn't a region of massive chemical farming. I've not seen a skylark or heard a cuckoo for years. Or a purple-sequinned sardonic warbler, come to think of it. Red kites are becoming quite common, on the other hand …

Every year swallows turn up after their unbelievably long journey from Africa. It's a lovely thing, to watch them swirl and swoop across the evening sky, grabbing flies on the wing. How quickly their minds must work. Such little brains, but they can dodge and kink at speeds a thousand times beyond the range of human technology. And how do they know their precise way back to our Blue Barn? Perhaps 'magnetism' might guide them north–south, but what about east–west?

When breeding is over in late summer, they line up on the phone wires, spelling out a tiny Bach fugue, and sing to each other like a row of Edwardian typewriters: ticka-chuckla-ticka-ticka, at very high speed, with an occasional a trrrrrrp, as the carriage returns.

Then one morning they're gone. Back to Africa, even the little babies. What a wonder.

Their greatest enemy seems to be magpies. One year these thugs wrecked every swallow nest in the barn and presumably ate all the young, or just stamped on them for fun. I don't like magpies, but they tell me that they don't actually reduce the number of songbirds. I don't understand this, I must say. No doubt the magpies' numbers rise and fall according to the inexorable rules of the food chain, but there do seem to be a lot of them about these days. One day I saw four perched cheek by jowl on a willow branch: a gang of drunken priests at a medieval conclave, shouting and bawling and jostling for position, all gob and elbows.

I once heard a lot of squawking from the yard and watched in bemusement as a magpie chased a *cat* away and up the drive, wings outstretched, tonsils ditto.

On another occasion I was sat on the patio, reading quietly. Suddenly something ran over my foot. It was a polecat dragging a baby rabbit into the shelter of the rockery, hotly pursued by ... a magpie, bawling and carrying on, as one who has been wronged. But I can't imagine the bird catching a rabbit, so I assume it was trying to thieve it.

I think 'unlovely' is a good word for magpies. Beautiful iridescent colours on their wings, but that's about it. Ungainly flight; ugly voice; thuggish behaviour. The SS of the bird world.

* * *

We have an old Burco boiler next to the front door, full of periwinkle plants. One spring day we saw a blackbird fly into the thicket of stems and then out again. Over the next few days there was a lot more toing and froing.* Nesting!

* Should these words have hyphens, I wonder? 'Doing' doesn't need one. Nor does 'going' but somehow 'toing' looks as if it might rhyme with 'boing' and mean 'the ringing sound made by tapping a tiny Chinese gong or a footballer's skull'. 'Froing', I think, must be the sound made by a percussionist riffling a rack of miniature tubular bells.

Obviously we were going to take great care not to frighten the new mum, so we re-routed our access to the house for several weeks, and hung a note on the gate to keep visitors away.

One day I peeped into the tangle, and lady blackbird glared back at me like a librarian on the Overdue Books counter. 'Sorry ma'am.'

Once the eggs had hatched and the feeding frenzy was under way, I waited for mum to dash out for a spot of exercise and a quick facial, and took a photo of the nest. Four chicks, all with that look of wide-eyed disgruntlement that every fledgling seems to have. I hope the flash didn't upset them.

They eventually all grew up and left home, although we missed the actual moment. A week or two later we found one of them dead in the yard. One of the three feroid cats that patrol our patch must have got him. Cats, eh? Another youngster seems to have broken his neck flying into a window. Windows, eh?

PS: the following year the blackbirds returned, but seemed not to take any notice of last year's periwinkle nest even though it was clearly up for vacant possession.

Instead they started rummaging about below it, in the lower recesses of the tangled stems. Then they stopped and disappeared again. It seemed a pity to waste such a splendid structure so I put a little yellow plastic duck in the old nest.

Two days later I was sat at my keyboard and saw a black flash pass the window, heading straight for the Burco at very high speed. I went to check. Had they decided to rebuild after all? What I found was that the original nest was now empty. The plastic duck had been bodily turfed out and was lying on the concrete, yards away. Quite an achievement.

They did eventually nest – in the rockery, of all places, just two feet above ground level, and successfully raised a brood.

I'll take the yellow duck out of the old nest again in time for next spring, just in case. Meanwhile, it provides a nice cheery face in the shrubbery to welcome us home.

A healer

Life continued at its steady pace, each season bringing fresh challenges as it has always done since farming began. Farmers are adaptable people, and take little for granted, we'd come to realise. Close to Nature equals close to Reality.

We sowed, weeded, cropped and ploughed, year after year. Practice and thought brought steady improvement to our efficiency, but we still had the recurring problem of whether I would be fit enough to reap what I might have sown.

How could we best cope with this indeterminate problem? I was neither well nor unwell; neither fish nor fowl; rather like the Platte River was to the pioneers in the Wild West: 'too thick to drink; too thin to plow'.*

We kept on trying to find a cure, and I began to visit another healer. Sally was clearly a genuine soul, and affordable into the bargain. And as she lived in Carmarthen, we could drop by after trips to town.

Did she help? Well, I *think* she did. Certainly, something odd happened as she laid hands on my shoulders. Her hands would occasionally vibrate in a strange manner. I've tried to

---

* I don't know about 'too thin', actually. Long periods of inactivity meant I was putting on far too much weight. Definitely 'too thick', though.

duplicate the movement, but can't. As far as I'm concerned, some sort of power was being transmitted.

After many visits the condition may have been alleviated, but, no, I wasn't cured. It's so hard to judge the effectiveness of any treatment, orthodox or otherwise, especially for something as peculiar as ME.

Anne continued to dose me with various pills, from vitamins E and C to ginseng and echinacea.

* * *

Paddy got his degree and set off into the big wide world, computer-wizardry buzzing round his head. Where would it lead him? 'To happiness', we hoped. Despite having been effectively fatherless throughout his teen years, he was a well-balanced young man, so he would be fine, we were sure of that. No doubt Life would throw challenges at him, because after all, that is what Life's *for*, isn't it? – but we were sure that he had the inner resources to cope, which is what matters. He would grow, as he ought.

– 29 –
Winemaking

Dad started making wine way back in the 1950s, when hydrometers were hammered together from 15lb of riveted brass and were unobtainable anyway, so knowing when his brews had finished fermenting was pretty much a hit and miss affair. An occasional barrel was bottled prematurely. Thus, forays to the pantry meant braving a gas-powered broadside to my tender young legs. I walked with a limp for weeks after the cork from a bottle of Parsnip '56 drilled into the side of my kneecap, and I retain the scar to this day.

His expertise improved enormously when hydrometers became affordable, but, curiously, he never seemed to use one for beer.

We arrived at his house one hot August afternoon.

'Fancy a pint?'

'Not half.'

Dad began loosening the top of a quart bottle. Before he'd managed a full half-twist, the bottle emitted a whistling shriek that started somewhere near Jupiter and gradually increased to a vampire scream as he twisted a little more. The jet of foam hit the nine-foot-high ceiling with such force that within five seconds there remained only an inch of blistering liquor in the bottle.

We escorted the remaining eleven grenades gingerly outside and defused them one by one. I've never been so scared. I have a photograph of foam *still* travelling at forty-five degrees, six feet to the right of one microscopically loosened screw-top. Since then I've had the profoundest respect for anaerobic activities and always use a hydrometer.

* * *

There aren't any fixed rules for wine recipes. I reckon on using 3lb of any sort of fruit per gallon, as a rule of thumb. More will give stronger flavour; less goes further. Add about 2lb of sugar for a half-decent tably/sideboardy sort of wine.*

Some yeast will tolerate up to twenty per cent alcohol, but if you fancy trying it I recommend beefing up the fruit as well, otherwise the resultant tincture makes your eyes water uncontrollably, moments before your shins give way in the middle. The extra fruit reduces the eye-watering by up to four per cent. Thus you will at least have half a chance of seeing the floor coming up to meet you, so you can try fending it off with your hat.

It takes about twice as long to make five gallons as it does to make one, so I tend to make five or even ten gallons at a go. It is a military procedure. All surfaces are cleared; all plastic dustbins, muslin, strainers, spoons, stirrers, tubes and fermentation vessels are sterilised with sulphite; then begins the process of scalding and mashing and being generally and Gothically beastly to 30–50lb of fruits assorted. When it's

* Incidentally, Continental winemakers call sugar-adding 'chaptalisation'; UK Trading Standards, on the other hand, call it 'cheating'. Take your pick, possibly bearing in mind that 'chateau' is the French for 'cat water'.

all cooled down, in goes the yeast, along with odd bits of nutrients, and I stand well back for a day or so. The liquor is then strained, unlike the quality of mercy, if Shakespeare is to be believed, and then diluted. I now add most of the sugar, dissolved in water. This is a tiresome business, as everything gets incredibly sticky. For God's sake keep the dog out of the way.

After that, it's mainly a matter of listening to the music of the fermentation locks: the constant glop and glup of escaping CO_2. It's good for your house plants.

I once had thirty-six gallons on the brew at once, mainly in those five-gallon polythene cubes from which off-licences used to sell British sherry by the glass-slipperful to desperate housewives. The fruit I used was mainly elderberry from local hedgerows, and peaches and plums rejected by the wholesaler (mainly in perfectly good condition, needless to say). All excellent cuvées, graded according to future use:

Grade	1	for	guests,
	2		dessert,
	3		table,
	4		recreational use,
	5		emergency,
	6		brush cleaner,
	7		drain scourer,
	8		poisoning rodents,
	9		visiting wine snobs.*

* Served from a fancy bottle with a misleading label.

The elderberry went down particularly well with the coalman. Poor lad, he wandered into our yard to tell me he'd stuck his overloaded pick-up into our ditch. I towed him out with the Volvo and consoled him with a dose of Elderberry Kneetrembler. This consoled him so well that he had another, then off he went, and got stuck in the ditch again. But this time he didn't mind.

If I'm lazy, I'll knock up a quick gallon from a bag of sugar and a carton of grape juice, but it's a bit on the vapid side: think the Moonlight Sonata on the ukulele.

I've tried all the usual fruits and veg, as well as curios like bananas, peapods, rose petals, and whelks (no, just kidding). They're not really worth bothering with. The peapod tasted of nothing; the rose petal was as thin as pity and smelled faintly of … nothing; and the banana produced an inch-thick band of clear grey liquor across the *middle* of a demijohn of brain-coloured slurry that was impossible to either siphon or strain, and which stripped the fur off your tongue and the cat. Yes, 'dryish'. It got mixed with a gallon of over-sticky currant wine, and eventually turned out a treat. In winemaking, two wrongs frequently make a right, I've discovered.

Currants are very good for adding body, but be warned: we were once unaccountably plagued with clouds of fruit flies in the kitchen, and used up dozens of fly-papers. Eventually we discovered an unfinished bag of currants at the back of a corner cupboard, seething …

My next brew will use the 40lb of sloes waiting in the freezer, plus whatever grapes we get from the polytunnels; and a few elderberries to boot.

I'll leave you with this hard-learned maxim: 'Never Give a Builder Strong Drink'.

And don't forget: Time flies like an arrow, but fruit flies like bananas.*

* * *

We did have plans for starting a little vineyard in a sheltered corner, but the most suitable patch of land was genuinely unfenceable, and rabbits would have killed every plant overnight. So it was the tunnels or nothing.

In 1994 we planted Poly 1 with a Black Hamburgh, as grown at Hampton Court; a year or two later we planted Poly 2 with a variety recommended by a specialist grower for our local conditions. It's called New York Muscat.**

The vine is a truly remarkable plant. The stem looks gnarled and scabby and, frankly, dead, but every spring it spurts out scores of yards of shoots and tendrils and hundreds of huge leaves. We never water our vines, but every year they fatten up dozens of bunches of fruit.

In a good year, the Hamburgh gives twenty kilos of grapes off a thirty-foot run. These tend to be small and rather pippy, and no doubt we would get a 'better' crop if we pruned more assiduously. Then we'd get fewer but bigger grapes. But for eating as nibbles, or for juicing (just over a kilo gives just under a litre) or wine-making, size doesn't really matter.

What does matter is sweetness. Hot sunshine is the key, and West Wales is at a slight disadvantage here. Some years the grapes are just a bit sour. Still good enough for chucking at the

---

* And currants.

** Yes: 'New York'. Not obvious, is it ...

sheep though. And if fermented, they make a perfectly acceptable Grade 7 or 8.

The exotic Muscat produces sweeter grapes … and oh! … that aromatic flavour. No doubt I could make a remarkable wine from them, but the crop is not nearly as heavy as we get from the triffid in Poly 1, so the five kilos or so get eaten, deliciously, straight off the vine. Mmmm …

We do make some effort at pruning, and April loves the trimmings. We make a bit of a thing of it, actually. On Pruning Day I dash into my little white ra-ra number and pom-pom slippers, clip on the stew-strainer moustache, and thrash about on my bouzouki while Anne twirls round and round with a rush basket held aloft, and posts the clippings into April's waiting maw, one sunsoaked sprig at a time. Satyrs prance and gambol in a rhythmic circle nearby, occasionally pausing to blow a chorus of 'El Condor Pasa' on their vine-wood syrinxes. I've never really understood why they do this, but there you are.

April tried joining in once, until the intricate cross-steps proved just too much for her simple little brain and she ended up in something of a knot, puffed out and leaning up against the hawthorn bush. We, of course, in our Dionysian ecstasy, merely laughed at her, and took another swig of grape-tinted turps.

In the autumn every Muscat leaf becomes a unique kaleidoscope of reds and greens and yellows and purples: each one a magnificent work of art. Tate Modern? … a refuge for the lost and the lonely, I'm afraid.

I was surprised at how easy it is to propagate vines. You just clip off a few prunings and stuff them into a pot of soil and compost. They will almost certainly grow if you keep them watered and out of the frost. I'm sure you could grow a bunch or two of delectable Muscat grapes on a carefully pruned plant in a tub on a sheltered patio; possibly even on a windowsill.

* * *

Home brewing and pets rarely mix well. Remember the 'drunken pigs'?

Another lady told me about the day her husband rang her at the office. This could only mean some sort of mishap concerning Magnus. Magnus was her, as opposed to their, blue-eyed, cross-eyed, very white cat.

When she'd left for work there had been a dozen bottles of elderberry wine maturing on the top shelf in the kitchen. Now most of them were splattered all over the floor, the furnishings, and the cream paintwork. Purple everywhere. Also a purple cat, flat out on the floor, blind drunk.

This was a pity, not only on account of the global mess and the wasted bottles of wine, but also because Magnus had only just regained his pristine colouring after falling into a bucket of fermenting damsons.

Cats, eh?

Cait's art and the Golden Rule

Cait had grown out of her days of dens and endless pets, and was growing up faster than we could keep up with. She did well at GCSE and went on to A levels. She said she'd be doing four, including her favourite subject of 'art'.* In our day, three

* History, sociology and *physics*, since you ask.

was plenty, thank you. But it seems competition was rising fast, so four it must be. She got her head down and worked, worked, worked.

The art acted as a sort of safety-valve of creativity for when the academic stuff got a bit much.

Cait's always had an uncanny aptitude for 3D work. I made a sort of stool/poof thing for Anne's fortieth birthday. It was designed to look like a mushroom, with a red domed and padded top, and a brown hollow body with a little door drawn on it, numbered '39a'. It was to be for storing needles and cotton and so forth. Paddy designed and made a decorative fabric caterpillar which he tacked round the side of the 'stem', and to lie on the cushioned top, Cait made a free-standing butterfly from wire and fabric and wood. She was six at the time. I couldn't make a better one myself, even now.

Her speciality is 'big'. Thus we now have a psychedelically coloured rhino, yellow, red, pink, purple et al., slightly under 4ft x 3ft x 3ft in size, supporting trays of plants in the conservatory; and a full-sized model of a moderately famous French art critic in the hall, sat at a desk, in a top hat, on a dais, and wearing a pair of my shoes, which I'd quite like back, actually.

The trouble with 'big' is that it takes up a lot of room. I encouraged her to insert a pencil sharpener carefully under the rhino's tail and to try selling it to a bank or insurance house as a bit of decorative but practical community art. No dice.

* * *

My off-days were filled with reading and studying now. After discovering the flaw at the heart of Scientific

Materialism (SfaS) I took a look at our three 'local' religions, to see if they could offer any guidance in my search for Truth in any way. They all had merits in that they all agreed there must be an Absolute Cause behind Life, the Universe and Everything, but all three showed similar flaws, to my mind.

The problem with Christianity was that it has no intellectual basis. Its driving principle is 'have faith'. My problem with this is that 'faith' alone does not *explain* anything, and can even lead to disaster: the Hitler Youth 'had faith', and much good it did them and the millions they went on to murder. And weren't the Spanish Inquisition's human bonfires known as autos-da-fé? ... 'acts of faith'?

My logical sort of mind needed something more intellectually satisfying than unreliable 'faith', but where might I find it?

Islam, too, is faith-based, and not strong on rational explanation; so is Judaism, as far as I can see. Too many opinions; not enough facts.

Yogi Ramacharaka had impressed me with his insistence that the reader should not take anything he said as 'gospel' (which made a refreshing change), but should test it himself for logic and inner consistency. And to my eyes it *was* logical and consistent. Here was something I could build on.

From Yoga I dipped into Buddhism, and other Indian philosophies. They all had one essential in common, which was that they appealed more to the mind than to the emotions.

But of course the emotion-based 'desert religions', Judeo-Christo-Islam, all had millions of thoughtful, 'mind-led' adherents as well as all their 'emotion-led' followers. And without a doubt the basic Christian ethic was impeccable, as no doubt were those of the others.* So maybe there was something I'd missed.

And yes, there was. It came as a revelation (!) when I realised that all religions are Januses: they have two faces. The outer public face: the *exo*teric; and the inner hidden face: the *eso*teric.

Briefly: the outer faces, the ones we see in the hugely varied rituals and processions and smells and bells, and so forth, are the ones that cause so much trouble in the world: '*We* face south-west to splatter the holy ointment over the Brick of Life, but *you* … you face due south … you … you *heretic*, you! Death! Death!! Death to the heretic!' And so on, down the ages. *The Life of Brian* says it all, brilliantly and definitively: 'Follow the gourd!' 'No ... follow the sandal!'

But if you look at the inner face of all these squabbling sects, you find something remarkable: they are all so similar as to be virtually identical. The *exo*teric forms of Hinduism show us millions of people rushing into the Ganges, or making offerings to rats or snakes, or multi-limbed symbolic fantasies, but the *eso*teric side, the side exemplified by Yogic philosophy, is entirely different. Not a snake in sight. Just reasoned and logical argument, suggestion and discussion.

* I was astonished to discover that the 'Golden Rule' of *'treat others as you'd like them to treat you'* is the *central* requirement of, wait for it: Islam, Judaism, Christianity (all varieties and sub-varieties), Buddhism, Sikhism, Jainism, Hinduism, Confucianism, Taoism, Zoroastrianism, and … Paganism. What are we to make of this extraordinary coincidence (if it is a coincidence, of course)?

Similarly with Islam: on the outside, we see huge pilgrimages and stringent dress requirements, but on the inner side, the Sufi side, we see something else, remarkably similar to that which the Yogis speak of. Judaism has its music and feasts on the outside; and an intellectual programme on the inside: the (real, not the rip-off pop) Kabbala. Buddhism is almost all intellectual; indeed many Buddhists say it isn't a religion at all, merely an explanation.

And what of Christianity? The nearest thing to an intellectual esoteric side is the writing of the mystics, but they are really more emotional than intellectual. We in Western Europe are ill-served for reasonable, rational, cosmologies. I guess that's why Buddhist groups are burgeoning while the traditional Church continues to decline: as we become better educated, so we require more satisfying mental food.

* * *

'So what?' I hear you cry.

Well, here's what: I now realised why science rejected religion in such an absolute way. It became *exasperated* with Christianity's endless vague flummery and pontificating (not a religion of the intellect, you see ...) as hopelessly irrational, and thus rejected the whole kit and caboodle, Absolute Cause and all.* This matters to us all because we've grown to respect

---

* Which unfortunately meant that science now had to replace the logical 'Absolute Cause' with 'Spontaneous Self-Creation': something of an embarrassment for any logical thinker, especially a scientist. In fact, in requiring Spontaneous Self-Creation to occur, for not just the Universe, but for Life, Mind, and Consciousness, as well, you might argue that science requires more miracles than religion ever did (SfaS).

science so much for all the help it has been to us, that we have even adopted its core dogma ... *that there is no cause or purpose to the universe and the world.* What's more, a lot of people have logically extended this nihilistic assertion to the conclusion that there is no purpose to themselves either. Despair (...depression, drug addiction, suicide, shopaholism, mindless consumerism, etc...) is only a step away.

Big social stuff derives from this rejection of the Absolute Cause. That's the *real* 'So what?'

There's more: when science rejected religion as being irrational, it also rejected everything else it deemed to be 'supernatural', by which it meant anything it could not measure. This is how it came to utterly reject, *in principle*, everything from spooks to dowsing, and from telepathy to homeopathy. 'Susceptibility to measurement' is the key. But what *are* the units to measure, say, dowsing with? And if you ignore the issue entirely, how are you ever going to discover these units?

Meanwhile, no scientist would deny that he himself is alive, thoughtful, and conscious. But he has never been able to apply any sort of measurement to these realities. Where does that leave him? Dead, thoughtless and unconscious? Of course not. So perhaps 'susceptible to measurement' is not the measure of all things after all.

Or maybe we just need to try a bit harder? We could start by abandoning *all* dogma.

Just as a matter of interest, where did all this Churchy flummery and pontificating come from? History shows us that 'religion', in its many forms, has become endlessly debased and diffused down the centuries, drifting further and further away from its core principle of the Golden Rule. Take

Christianity as an example: nowhere in the Bible will you find Jesus commenting on how many fingers you should make the sign of the cross with, or whether or not you should eat fish, or cover your head (or not) in a church, or burn people at the stake. ALL that, and just about everything else that passes as 'Christianity' has been tacked on by priests and popes and presbyters. Jesus' summary of religion was:

(a) accept that there is an Absolute Cause, who/which is smarter than you are
(b) accept that this Cause is benevolent
(c) act accordingly towards It and each other (the Golden Rule ...).

That's IT. Jesus wasn't even 'a Christian'. He just came 'with a sword', as he said, to hack through all the stupid ritualisation that had built up around the original simple ethical revelations to Moses.

There have been many other reformers and Reformations, from Akhenaton and the Buddha onwards. There will no doubt be many more. We learn very slowly, it seems!

* * *

Which brings us up to 1995. Our thirteenth year on the farm.

– 30 –
Creativity

We came to smallholding from a background in education, a brainwork profession. But we've both used our brainpower far more on the farm than we did in education.

Odd? Not really, because 'school-brain' is actually rather limited: learning stuff, trotting it out, checking the feedback. 'Farm-brain', on the other hand, requires all kinds of other skills. You have to plan against not just one or two factors, like exam syllabi and lesson length, but against a host of variables, some of which you can't control at all. You need to calculate which crops to grow, which varieties, in what quantities, and at what intervals. You need to do all this with specific markets in mind at the time of sowing without knowing whether your crop will actually be in demand when you've picked it. You need to ensure that you follow a suitable rotation, succeeding deep rooters with shallow, nitrogen-givers with nitrogen-takers, and make sure crop A does not over-shade crop B. Also you need to have a green manure of suitable qualities available to follow each harvest; and you need to be sure to spread your harvests throughout the season, so you don't have everything cropping at once. And you never know what the weather has in mind or which of God's creatures will see fit to share, not to mention devastate, your crop with/for you. You might wish to invent ways of limiting their predations.

You also need to learn some rudiments of animal medicine and when to apply what to where and how. And then there's a multiple modicum of technical stuff on the machinery and building fronts.

Above all, you need to learn how to improvise, bodge, adapt and invent: you must learn to become *practically* creative in many ways you may not have been before.

When we first arrived I remember taking instructions with laughable literalness. If it said 'smash with a hammer', I would not consider using the back of the axe I was holding in my hand. Embarrassing to relate, but quite true. These days I look for the *spirit* of the instruction: 'apply lots of force, with ohh, I don't know … a hammer? Boot? A stale Eccles cake?'

Self-harmer Wasting Hospital Time
With 'Impacted Raisins' In Hand
'Already Chopped His Finger Off' Claim
See pp. 2–3 For Three-colour 36-piece Jigsaw Offer
'Hippie Chops Both His Hands Off' ONLY £8.95
Two Free Mugs With Every Purchase

We soon began to revel in inventiveness. It was fun, rewarding, and saved money we didn't have. We needed bungees to hold the polytunnel doors on with, so we made a dozen from an old bike innertube, a length of baler twine, and some hooks, roughly bent from a bit of damaged pigmesh. £10 saved for half an hour's work. Not a bad rate of pay, even today.

We needed to regularise our plant spacing. A quincunxed

bed* grows twice as much crop as the same landspace using normal rows. But how to lay out this triangulated lattice? The answer was to make a multi-dibber from a 2ft x 6in slab of scrap plywood and a dozen hazel pegs cut from the hedge, screwed in as two off-set rows. You lay the edge along your field-line (200ft of baler twine, naturally) and off you go. With a cunning notch at the leading edge to ensure accurate registration, you can dib 1600 holes almost as fast as you can walk up the field. You then repeat, until the whole bed is marked out: 7800 holes prepared in twenty minutes. Easy.

* The five-spot on a di(c)e is a quincunx. Imagine a line of five-spots, 200ft long … you end up with three 200ft rows, each spot equidistant from its neighbours, do you not? Optimum spacing …

The multi-dibber was such a success that I made a bigger one for wider spacings, with a taller handle on the back that speeded things up even more; and eventually the Mega-Dibber, which was a step too far, being a full-sized door with dozens of pegs on it, which you hurled to the floor and then jumped on. Yes, I know … too big, wasn't it? Not too heavy, although it was heavy enough on a windy day, but too rigid: you couldn't get enough pressure on each peg on uneven soil. I eventually sold it to a Medieval enthusiast wearing studded wrist-bands and a matching collar. I forget what I told him it probably was.

Old audio and video tapes, stretched out to flap and buzz between bamboos, make useful bird scarers; fifty two-inch lengths of ruptured hosepipe will save you £25 on electric fence insulators; one old orange-juice tetrapak makes two excellent new plantpots; you'll never need to buy another carillon or wind-chime if you've got enough bean tins and oil cans; bits of field drain make excellent tree collars. I love it.

I'm sure much of the general urban malaise, especially amongst kids, is the lack of opportunity for inventiveness and genuine creativity. Our society's materialist philosophy has drained away all our instinct to create, and has instead sucked us all, via relentless advertising, into the role of passive 'consumer'. What a deathly and defeated word that is.

'And what did you do with your life, Mr Jones?'

'Well, St Peter, I was a consumer.'

'Were you *really* …?'

Pallets: wonderful for making compost heaps, gates, sheep hurdles, interior shed walls, cheese presses, tables, and chicken huts. Also love spoons, pattens, and novelty bookmarks.

Old tyres come in handy as silage weights and gate-stops,

and for making swings, stepping stones, and emergency shoe soles. Lorry tyres are handy for creating bolt holes in hedges that let sheep through, but not cows. When cut across the diameter, a half tyre makes a perfect blade guard for a circular saw. And a tyre on the wheelrim, cut round the circumference and 'popped' over, makes a brilliantly inventive flowerpot/planter or travelling commode. Tyres are also useful as buffers against clumsy driving, and ear-rings. No, not ear-rings.

Baler twine is useful for everything from surgical suturing to garrotting pigs, should you feel inclined. Anne made a very attractive three-colour plaited strap for an old school satchel full of hand-tools that gets hauled everywhere with me in the weeding season.

I had lots of fun making a garlic-dryer from a set of steel shelves panelled with insulated hardboard front and back, but with the sides interconnected by semi-cylinders of 2mm polystyrene 'lining paper'. The minced garlic lay on trays on the shelves, and a small radiator warmed it from below. The warm air passed from shelf to shelf via the polystyrene guides. It even worked, after a fashion. The garlic came out as hard as shredded topaz and incredibly potent if you could only chisel it off the trays (goggles essential).

My neighbour makes huge stoves for his huge shed from pairs of oil-drums. He once made a stove from an empty liquid gas bottle. Why didn't it explode as soon as the acetylene torch touched it?*

* No, it's no good looking here. You'll have to work it out for yourself. That's the fun of it, after all. But *please* don't try it until you're certain you've cracked the secret. In fact, please don't try it at all, as it is almost certainly illegal, never mind ridiculously dangerous.

* * *

On the spacing front, we also built up a stock of 8ft bamboos, marked off with masking tape at various intervals. Planting swedes? Then pick the one taped every 8in. Spring cabbages? Pick the 12in one. If we were feeling dangerously continental, we might tape at intervals of 203.2mm, 304.8mm etc., and perhaps use different coloured tape to indicate different spacings on the one pole. A gaily decorated pole is a sure-fire ice-breaker at Work Parties and thés dansants.

* * *

Despite all our efforts, it gradually became clear that our days as a productive commercial mini-farm were numbered. The ME just refused to go away. True, it abated, and would sometimes disappear for a whole week, but the farm economy needed me to be working fulltime and predictably, and that was that. Anne worked wonders, but there was a limit.

It was a question of 'when', not 'whether' ...

All change

The crunch came in 1997. That was quite a year for Britain, as you might remember, and particularly memorable for us.

In May we had the election that finally excised the Tories after 176 years in power. Surely now things could only get better? Surely we would see the divisive loadsamoney culture of selfishness and greed replaced by something more human, civil, and intelligent? Perhaps we dared to be optimistic?

Then in July, Mum died. She was 83 and had been ill for

a long time with some sort of lymphoma, but it still came as a shock, as all deaths do. It's that sudden and absolute emptiness: a room filled with an expanding absence; personal knick-knacks and effects at every turn: items of handwriting, knitted clothing, worn handles. It hit Dad very hard, naturally. To cope, he started to teach himself Ancient Greek.

In August Princess Di was killed, and fifty million flowers bloomed into our collective memory.

In October, Cait left home. She'd passed all her A levels with splendid grades and went off to Oxford to study Philosophy, Politics and Economics. Not my idea of fun, but Cait seemed to be looking forward to it.

She pointed out that, after all, it wasn't the first time she'd left home. It seems that when she was eight she had run away to one of her dens, because nobody would spare the time to help her make a fishing rod. My irrefutable defence is that I have no recollection of this request; and anyway, it was the middle of summer and, quite frankly, picking runner beans was more immediately important than helping Number One Daughter land a marlin, or stickleback, or whatever.

Anyway, she had then had an argument with Paddy over something, which was the last straw. In true Beano style, she tied a red spotted hanky full of vital supplies (biscuits, teddy-bear, etc.) to the end of a bamboo, and set off. She even took newspapers to sleep on and under, and decided on the den halfway up the drive (out of the twenty or so others she had) because it had good shelter under a hazel tree, which would be good for nuts, and also had a stream running through the corner of it which would supply all her water needs.

Clearly Oxford would hold no terrors for such a competent young person. We all smiled bravely and waved the train goodbye.

Now we original six were down to three. Our world had changed enormously.

* * *

In fact, Cait's leaving was just the final wisp of straw. For years we had been adjusting our production and methods to fit our health circumstances. First we cut back on suckler calves, then on calving; then on milking. We cut down on courgettes and cabbages; sold some sheep; gave up on several crops completely. Bit by bit, we were retreating, and we knew it. We stopped supplying a couple of shops with veg, and then the wholesalers. Anne saw the writing on the wall more clearly than I did. After all, it was she who was doing all the extra work.

And for what? We hadn't set up the smallholding so we could become medieval serfs, slaving for the benefit of the middlemen and supermarkets. We had set out to enjoy working together, to produce as much of our own food as we reasonably could, and a surplus to sell as a cash crop. Now we couldn't work together, and Anne alone simply could not do enough to keep us afloat. I was already going onto Benefit over the winters, as our modest but adequate income slumped beneath the sustainability line. What next?

The realisation hit me in the autumn, when we came to harvest the spuds. Anne had originally wanted to plant just 50ft, which would give us a small but helpful crop. I, however, in a bout of healthy springtime optimism, thought we should go for 400ft, and do the job properly. So we did. They grew nicely.

But came the cropping season, and I was floored again, slumped and useless, so Anne had to do all the lifting herself. I wandered out of the back door at one point, and saw her on the field. She looked weary, but she didn't stop. Fork in, tilt, bend, heave and scrabble, find the tubers, bag them, fork out. On to the next. There was no vigour or joy in it. Just duty and will. And another 350ft to go.

I felt impotent and ashamed. This was all my fault. Through my own lack of sensible foresight I had doomed the already overloaded love of my life to hours and hours of extra hard labour. Things had to change.

* * *

As for any parents, now that the nest was empty it was time to reassess everything. I apologised for the potato fiasco, but it wasn't a case of apology. It was a case of facing facts.

One evening in December we sat in the sitting room with a bottle of peach wine and looked at each other. Neither of us wanted to be the one to say it. I can't even remember now who it was who did the daring thing, but I can still feel it. We looked each other in the eye and tears welled up and rolled down our cheeks. 'We'll have to close.' Nods. Sniffs. A forced smile. More sniffs. More nods. 'Yes.' We raised our glasses to each other, but had no toast to make. Too many emotions. The gesture was enough: 'Here's to you; here's to us.'

* * *

I wish I could say we felt better for having made the decision, but we didn't. Some weeks later I asked Anne how she was feeling about it all. 'Lost, lonely, and in limbo,' she replied.

She was lonely because I was no company, and had not been for years; she'd had to take on everything from taxing the car and keeping the insurance paid to looking after all the animals and the kids. *All* the planning of crops and everything else had fallen onto her; *all* the responsibilities; *all* the burden. I was simply too unreliable and forgetful to be trusted with anything. Now that Cait had left, she would be even lonelier.

She was lost because the Plan had gone: in cliché terms The Dream Was Over, although it had never been a dream in our eyes. We'd gone into it with open eyes, and knew it would be hard work. We simply could not have foreseen the ME-Joker in the pack.

And she was in limbo because she could see no future, except eking out an existence on Benefit. I was literally unemployable as I couldn't be relied upon from one hour to the next. Anne couldn't get a full-time job because we still had April and a few sheep and the farm needed servicing even if it wasn't a commercial outfit. And anyway, who would want to employ her? And for what?

The winter passed in tones of grey. After fifteen years of trying, it was hard to admit defeat.

* * *

But as spring re-blossomed we began to come through it. Was it really a defeat? True, we could no longer support ourselves a hundred per cent, but we (mainly Anne, of course) could still grow our own veg in the polytunnels and on a garden patch. Our money needs were modest. We would get by.

In fact we could [illegible] retire! Several of our old college friends had already done so. So why not us?

It was this thought that caused the penny to drop for me. If my life really was a journey, then 'defeat' was out of the question. It was simply *change* that had occurred.

Looked at this way, nothing is a failure: merely experience, which will have future value, somehow. Excellent. And we had piled up a fair bit of experience over the past decade and a half.

We had learned a lot about animals, and life and death, and the complex inter-relationships between all living things. We had learned that food does not grow in a plastic bag, but must be nurtured and won from the soil, just as it always has been for thousands of years. It takes work and thought and is risky. We now realised in full that Man depends on his topsoil for his survival, and that he poisons it with excess chemicals at his immediate peril.

We had learned that lack of money need not mean poverty or misery if other things are in place: no debt, simple tastes, love, purpose, creativity.

We fully appreciated the importance of Green living: that every blast of energy used in industrial production means another shot of greenhouse gas released: that therefore every item we waste and replace, from half-used teabags to fitted kitchens, is all adding to the slow but accelerating destruction of the environment; that there is only *one* environment, and we all have a responsibility to do our bit for it, maybe by buying organic and/or Fair Trade, or by creatively making-do and mending a little more.

We had learned not to just understand these things, but to

put them into practice, so that we were living an integrated life. Our theories and philosophy were lived out every day. We practised the Three Green Rs as second nature: we Reduced use, Reused whatever we could, and then Recycled what we couldn't.

The result of our integrated life was that we were deep-down happy. Anne's lost and limbo feelings were not terminal. She would not want to swap her life for any other. Go back to the city? No thanks. We'd revisited Nottingham a few years back and felt like aliens. It took a while to work out why we felt that way, and we came to the unexpected conclusion that the place felt as if there was 'too much money' about. What a curious analysis, you might think. Or maybe you won't. I guess you are only reading this book because you already have half a feeling that maybe More Money is not the answer to Man's woes and ills and deep dissatisfaction.

We had also become incredibly efficient. We no longer took tools straight back to the shed or took washing straight upstairs. Instead, everything in transit would get left at various staging points en route, waiting for the next person passing to take it with them. We saved hours of wasted trudging every week that way.

We recycled and reused not just for Green reasons, but because it saved money, which meant that we could get by on a minimal amount, which meant that we never actually felt anxious about it. So my chair was falling to bits? So what? My shoes had cracks in? So what? I'll just use them in the dry. Big deal.

There were other positives too. We were glad we'd brought the kids up in the country, away from the manic materialism of the city, where having the wrong sort of squiggle on your pumps meant bullying and social exclusion. Out here they had learned at least a little of the values we had learned. They would never be wastrels or cynics. And they would always have the option of staying in the city as adults, or returning to the country, knowing what to expect.

Cait in particular has relished her childhood freedom to climb trees, make dens, roar about on dangerous trolleys, and keep pets. She also knew about burying pets. She and Edward seemed to have a funeral every other month: guinea pigs, rabbits, chickens, zebra finches, and the occasional puppy. A row of vertical slates in a flowerbed bear witness. John the Duck died aged four, and was a hard loss. She'd had him from an egg, along with his three brothers, Brian, Roger, and Freddie. All part of the learning curve.

So, *experience* then, and not wasted time. But what now? If the ME was not going to go away, what could I do with my life? And what could we do to minimise Anne's 'limbo'? 'Retirement' was never going to be a real option for us, we knew that. We would always need to be doing something positive and creative. What would the next step in our journey turn out to be?

My reading and thinking had brought me to a point where I realised that religion and science could actually be quite easily reconciled, despite the posturing, bafflement, and name-calling on both sides. Perhaps that discovery would somehow become relevant to our future. Perhaps I should write an article or something, and see what came of it?

Since 1989 I had been writing a column called *Smallholding Scene* for the HDRA magazines (latterly *The Organic Way*). They didn't pay me, and I didn't expect it. It was just something I *could* do, and if it amused a few people, that was payment enough. And I was finding that I was beginning to actually enjoy writing (well, on a good day, anyway).

Then one day in 1998 the editor rang for a chat about something and said that she got the occasional letter from readers, and that 'more than one' person had asked if I'd written anything else, like a book, for example.

'So how about it?' she asked. 'Why not write a book?'

'A book?' I thought. 'Proper *book*… instead of just odd scribblings? Hmm … why not?' After all, there would be *guaranteed* sales of 'more than one'. Who could resist?

Failed Hippie Claims: 'Going Forward, Singing All The Way'

Has Obviously Learned Nothing.

Ha!

Don't worry – I wasn't going to go without explaining how Alan the miner rescued the Co-op van.

The van was facing downhill. Alan threaded a rope through a slot in the outer of the double rear wheels and clove-hitched it to a spanner on the outside of the slot. He hauled the tensioned rope back uphill and hitched it to one of the big angle-iron tent pegs hammered deep into the ground in line with the back wheel. Then he did the same on the other side of the van.

Finally, he started the motor, stuck it into reverse and very slowly let the clutch go. The motor gradually winched the truck up the ropes …

A useful tip to remember next time you go to a school sports day.

****

A religion that's battered and wrecked
And a science that's become a new sect
Should take time to learn
They're two stones of a quern;
It's about time they tried to connect.

Yes, I do appreciate that if both stones were to 'connect' in a really serious way, that that would ruin the quern as a functioning device, but let's face it, who has the time to hand-grind corn these days? And anyway, the reunification of the rock might be a greater prize.